Social Knowledge Creation
in the Humanities

Iter Press

NEW TECHNOLOGIES IN MEDIEVAL AND RENAISSANCE STUDIES 7

SERIES EDITORS William R. Bowen and Raymond G. Siemens

MEDIEVAL AND RENAISSANCE
TEXTS AND STUDIES

VOLUME 526

Social Knowledge Creation in the Humanities: Volume 1

Edited by
Alyssa Arbuckle, University of Victoria
Aaron Mauro, Penn State University
Daniel Powell, King's College London

Iter Press
Toronto, Ontario
in collaboration with
ACMRS
(Arizona Center for Medieval and Renaissance Studies)
Tempe, Arizona
2017

Library of Congress Cataloguing-in-Publication Data

Names: Arbuckle, Alyssa, editor. | Mauro, Aaron, editor. | Powell, Daniel, editor.

Title: Social knowledge creation in the humanities / edited by Alyssa Arbuckle, Aaron Mauro, and Daniel Powell.

Description: Toronto, Ontario : Iter Press, 2017. | Series: New technologies in Medieval and Renaissance studies ; 7 | Series: Medieval and Renaissance texts and studies ; volume 526 | Published in collaboration with ACMRS (Arizona Center for Medieval and Renaissance Studies). | Includes bibliographical references.

Identifiers: LCCN 2017029560 (print) | LCCN 2017052721 (ebook) | ISBN 9780866987394 (ebook) | ISBN 9780866985833 (hardcover : alk. paper) | ISBN 9780866985833 (v. 1 : hardcover : alk. paper)

Subjects: LCSH: Humanities--Research--Methodology. | Humanities--Research--Data processing. | Group work in research. | Communication in learning and scholarship--Technological innovations.

Classification: LCC AZ186 (ebook) | LCC AZ186 .S67 2017 (print) | DDC 001.3072/1--dc23

LC record available at https://lccn.loc.gov/2017029560

Cover image:

Gouw, Tim. *Gastown, Vancouver, Canada*. Digital image. *Unsplash*. Web. 24 Apr. 2017 <https://unsplash.com/photos/NSFG5sJYZgQ>.

ISBN 978-0-86698-583-3

eISBN 978-0-86698-739-4

Contents

Tracing the Movement of Ideas: Social Knowledge Creation in the Humanities

Daniel Powell
King's College London
daniel.j.powell@kcl.ac.uk

Aaron Mauro
Penn State University
mauro@psu.edu

Alyssa Arbuckle
University of Victoria
alyssaa@uvic.ca

A Framework for Understanding Social Knowledge Creation

In *The Great Chain of Being: A Study of the History of an Idea* (1936), Arthur O. Lovejoy defines a subdiscipline commonly referred to as the *history of ideas*. Intending this area of inquiry to be a field that broke from the doctrinal lineages of particular philosophical schools and their attendant "isms," he presumes that the foundational logic underpinning most systems of thought can be roughly grouped. Lovejoy argues that once the surface level differences and contemporary concerns are removed, there are in fact a limited number of truly different systems of thought. Lovejoy's provocative work immediately garnered accusations of reductionist and relativist thinking. Believing in the primacy of the dialectic, his act of declaring a foundational moment of discursive analysis seemed to confirm his thinking.

As his title suggests, Lovejoy's particular brand of Christian-inflected Platonic dialectics posits that the metaphor for the history of ideas is a chain; whether we speak of Plato's allegory of the cave, or the Renaissance vision of a *great chain of being* linking all of creation, Lovejoy's use of dialectics relies on a sequential and ordered hierarchy of thought. It is grounded firmly in the communications technologies of its time (the 1930s), predating the advent of truly mass media in the postwar West, much less the deep interconnectivity of many twenty-first-century digital cultures. Lovejoy's thinking is hierarchical rather than networked. This appeal to hierarchies is no longer congruent with the nebulous networks of postmodern knowledge creation and transfer. Since the 1960s, the network has been touted as a metaphor for knowledge: Roland Barthes, for example, theorized the telephone as both a "cluster" and a "sequence" (1977, 94, 109). The scholarship contained in volumes 1 and 2 of *Social Knowledge Creation in the Humanities* is of a different

ISBN 978-0-86698-739-4 (online) ISBN 978-0-86698-583-3 (print)
New Technologies in Medieval and Renaissance Studies 7 (2017) 1–28

order. The unifying argument for the work is that the speed, ubiquity, and diversity of online platforms, tools, techniques, and interactions have generated and continue to inform distinct cultures of knowledge creation that champion Open Access (OA) publication, iterative process-based development of outputs, and broadly collaborative methods of authorship. Taken together, these sociotechnical changes have fundamentally reshaped how knowledge creation, exchange, and use occur in scholarly environments.

These methods—made possible by global information and communication networks—are simply not reducible to Lovejoy's conceptions, and often defy scholarly attempts to account for the actions of large groups in developing new ideas. Lovejoy's governing metaphor was the chain; many scholars today consider the network as more emblematic of contemporary information flows and exchanges. Sociology and psychology have both studied the ways in which complex systems of thought emerge from a network of individual actors (Social Life and Social Knowledge), and Social Network Analysis (SNA) is rapidly emerging as the ur-discourse for the twenty-first century. John Scott's *Social Network Analysis* has now entered its third edition since 1991, and corporations and governments routinely mine social media data to target consumers, citizens, competitors, and criminals. Regardless of whether these activities are fully sanctioned or understood by the public, there is a growing awareness about how technology and culture intersect in the age of the network. In academic scholarship, this means that the tools and techniques of networked data analysis have begun to link disciplines in new ways. Due in part to the prevalence of social media in everyday life, the sense that academics who work in collaborative teams are agents within a complex system is deeply familiar and even affirming. Understanding our agency and the ways in which technological systems shape our acts of knowledge creation, dissemination, and reception is a key component of this new disciplinary landscape, relevant both to the self-organization of the university and to its wider role in society at large.

Volumes 1 and 2 of *Social Knowledge Creation in the Humanities* address such widespread and systemic changes by bringing together diverse scholarship illustrative of this emergent ethos. These volumes extend John Maxwell's concept of open, agile scholarship:

> The embrace of the Open Web and its native agile approach opens up a space in which, rather than specifying outcomes and audiences in static terms, we allow for ongoing interpretation and intervention, where we allow our work, our ideas, our prototypes and models, to

live and evolve ongoingly, past our own imagining of their value. What kind of scholarly discourse will we see when the outputs of our work become not only accessible, but truly open: reviewable, revisable, reusable, remixable, by an unanticipated audience? This is the larger promise of truly open scholarship. (2015)

The collaborative research group Implementing New Knowledge Environments (INKE; inke.ca) has refined these ideas under the rubric of *open social scholarship*. Defined by INKE as "creating and disseminating research and research technologies to a broad, interdisciplinary audience of specialists and non-specialists in ways that are both accessible and significant," open social scholarship traces its roots to Open Access and open scholarship movements, as well as public-facing citizen scholarship and contemporary online practices ("Victoria Gathering 2017"). Taking open access as a starting point, and building on the work of open access advocates and organizations (Meadows 2015), open social scholarship asks: what happens after widespread access to research is granted? A founding premise of the field asserts that access to research data is not enough; research must also be presented in formats and in contexts that encourage public interaction and participation in a multitude of ways.

There is significant overlap between the terms open scholarship, social knowledge creation, and open social scholarship, and each phrasing emphasizes a different aspect of the same set of phenomena. Open social scholarship, at the moment, best captures the shared emphasis on collaborative processes and open, iterable output. Unified more by a shared ethos and set of practices than by any explicit disciplinary boundaries, this volume discusses open, networked, social knowledge creation and scholarship in terms of the local, the specific, and the limited, rather than a universally applicable and stringently demarcated criterion for any single term. Simply put, open social scholarship is less concerned with laboured conversations about definitions and disciplinary boundaries, and is more focused on adapting a design-based approach that is interested in communities of practice, pragmatic research and teaching, and evolving methods over time.

This introduction attempts to map commonalities among and between ways of *doing* open scholarship, charting the intersection of three broad and overlapping movements that deeply influence work in this arena: open access, crowdsourcing, and team-based collaboration in digital scholarship. It then moves to consider how these trends combine in humanities scholarship and communication, and ends with a contemplation of technologies as practice in the realm of knowledge work.

Open Access and Openness in Scholarly Work

Social knowledge creation is not possible without the open circulation of information. While this does not necessarily include personal data, an operational ecology of open knowledge work demands that academic resources, publications, and data be as open as possible. In scholarly publications especially, the logic of open scholarship has coalesced around the Open Access (OA) movement. OA publishing offers alternative funding models of editing and publishing services for universities that face budgetary pressures in an unstable global economy. Committing to OA research is both practical and crucial, as it allows those who fund research—i.e., taxpayers, primarily through the mechanisms of governmental funding bodies and institutions— to gain direct access to research output without barriers such as paywalls. OA also facilitates a core tenet of humanistic research: to share, analyze, circulate, and discuss findings widely.

In *The Access Principle: The Case for Open Access to Research and Scholarship,* John Willinsky defines this ethos as the "access principle," which demands that a "commitment to the value and quality of research carries with it a responsibility to extend the circulation of this work as far as possible, and ideally to all who are interested in it and all who might profit by it" (2006, 5). From a pragmatic standpoint, OA produces a more efficient knowledge production system, as students, researchers, and other stakeholders can learn from and build on their colleagues' insights. OA is often perceived as being at odds with traditional publishing models and scholarship practices. There are significant economic considerations in shifting from a system in which a handful of mainstream publishers control and profit from academic publishing to a system predicated on free and open access to research output—as Willinsky notes, access to research has never been simply "an open-and-shut case" (2006, *xxi*). Although counterintuitive to standard knowledge commercialization and profit models, open access development can provide a cost-effective commercial model by increasing the range of services that organizations are able to provide (Crow 2009; Kennison and Norberg 2014). These organizations may include small publishers and journals, as well as university presses. Increasing the diversity of media and methods used to convey scholarship to the public is a key characteristic of open social scholarship.

Creative Commons, the most well known creator-based licensor, identifies six primary economic benefits to an open model: (1) reduced production costs; (2) reduced transaction costs and legal uncertainty; (3) increased access to innovation and reduced marketing costs; (4) increased first mover advantage; (5) increased "opportunity benefits" and reputation; and (6) sustainability. Under the aegis of Creative Commons, many are now considering how to reinvent existing practices around knowledge work and its output—such as monographs and scholarly journals—within an OA framework, and with creativity and public engagement in mind (Guldi 2013, Saklofske 2016). In "Beyond Open Access to Open Publication and Open Scholarship" (2015), Maxwell calls for an evolved mode of online scholarly communication, which he considers as possessing "a conservative character, highly resistant to structural change." He criticizes academics for settling for a traditional and limited production system, and urges them to consider more agile, social, and flexible publication models that consider relevance and a reader's attention—and not just that of an assumed professional academic reader.

A culture of openness allows researchers to become more involved with each other and with other interested stakeholders, including those in the public and private sectors. We take it as a given that there is considerable value in researchers productively sharing their intellectual labour and outputs beyond what have historically been considered disciplinary, institutional, and social barriers in the service of a larger public good. Such openness is in line with increasing the impact of academic work as outlined by various funding and governmental bodies around the world; Research Councils UK articulates academic impact as "[t]he demonstrable contribution that excellent research makes to academic advances, across and within disciplines, including significant advances in understanding, methods, theory and application," paralleled by economic and social impact that measures "[t]he demonstrable contribution that excellent research makes to society and the economy" (Research Councils UK). The UK Research Excellence Framework defines impact as "an effect on, change or benefit to the economy, society, culture, public policy or services, health, the environment or quality of life, beyond academia" (2012). By working toward widespread acceptance of openly accessible scholarship and considering how to implement and inspire systems for open social scholarship, researchers can move beyond the stereotypical closed circle of their departments and fields, and move into a more open arena that invites and involves many different people. This increases measurable impact, and begins to shift scholarly norms of activity and behavior.

In a twentieth-century context, it made a certain amount of sense to have research disseminated through print publishers: physical production of textual materials was (and remains) expensive and infrastructure-intensive, in addition to the realities of managing the distribution of such materials. The physical needs of print were compounded by disciplines that required increasingly technical visualizations or images. The cost of high quality colour reproductions, for instance, is exceedingly high. For important research of this kind, companies that we often think of as traditional print publishers were the only entities capable of doing such work. In the late twentieth and early twenty-first century, the Internet has reduced these costs, and in so doing has radically restructured various parts of the production and dissemination process. On the content side, experts invested in their field of study have long undertaken the expensive human work of creating and revising research materials. Peer review is rarely factored into the cost of publication, as it is usually "free" labour from the publisher's perspective. Increasingly, graphic design, page layout, and grammar checking services can be attained on university campuses, and some web-based systems can automate this work. The technical need for large corporate publishers may no longer be present, yet academic culture has retained a regressive sense of prestige associated with the sort of publication and dissemination that those publishers offer.

The result of retaining a largely obsolete and irrational distribution system is an artificially high cost of access to original research. Paywalled research retrenches the perception of disengaged academics and a cultural elite who are uninterested in the real world. OA, by contrast, champions a breadth of knowledge stakeholders. Researchers, students, patients, cultural workers, governmental organizations, community members, and individual citizens are affected by the nostalgia for paper and fealty to notions of prestige and status. By contrast, digital media are likely the norm for these stakeholders when it comes to accessing music, literature, movies, and every other kind of content imaginable. While digitization is also the norm for scholarly publications, the price has increased well beyond the pace of inflation regardless of decreasing dissemination costs. Periodical costs have increased by over 250 per cent since the mid-1980s, and an annual subscription to the most expensive periodical in chemistry costs $4,488 (Bosch and Henderson 2015). Although the reality is that publishing costs are lower and subscription rates are increasing, academics still donate their time to these corporate publishers to review and edit the academic content that the publishers release. Furthermore, contemporary publishing costs impact what teachers can teach due to a lack of access or burdensome copyright law. Less wealthy countries can be similarly disenfranchised through nationally based corporate partnerships and currency devaluation. While there will always be a cost associated with publication, the current costs restricting public access are unrealistic and unnecessary. There must be a viable model for long-term, sustained, universal access to publicly funded research.

While the free-to-read argument is important, it is equally important that OA allows for large scale analysis. JSTOR's Data for Research portal is a good start for such access, but a truly OA world would not require restrictions.[1] Full text data mining across many journals and over many years can allow for patterns in research and new directions that are not currently possible because no single individual can read such large collections. HathiTrust's recently released Data API (application programming interface) is another noteworthy step toward a world in which large scale analysis is the norm, thereby increasing the speed and efficiency of research.[2] In some ways, scholarly publishing has always been collaborative and dependent on networks of people from around the world to share and improve research. The networks of relationships and the media-specific dependencies are changing. The relationship between paper and prestige is purely a cultural one, but the assumption that paper publications are more rigorous and more valuable will likely be present for years to come. After all, we might imagine our readers thinking, "Why would they have bothered to print them if they were not somehow special?" This aura of paper will fade in time, and the publicly facing and socially engaged researchers will continue to push the limits of academic cultural norms.

Crowdsourcing in Culture and Academia

Contemporary consumption of culture and information is deeply tied to a participatory sensibility (Jenkins 2006; Mahony and Stephansen 2016). Social media has helped to shape public perception about individual actions online, and we as individuals are becoming more and more comfortable in large, complex, and rapidly changing media environments. Those who actively and regularly use social media scan the flow of information and, at times, become a part of it. "Users" and social knowledge creators both create and consume online content. Social knowledge creation is deeply tied to the so-called "wisdom of the crowd." Crowdsourcing has become a symptom of broader collaborative structures made possible online. The term "crowdsourcing"

---

[1] http://about.jstor.org/service/data-for-research.

[2] https://www.hathitrust.org/data_api.

has long emphasized a primarily business-oriented activity, but in recent years it has grown to include an expansive branch of participatory, public-facing collaborative scholarship, and is now integral to debates about certain types of collaboratively produced knowledge in academia and beyond.

As a distinct term, crowdsourcing dates from a 2006 *Wired* article by Jeff Howe. In that piece, he details "The Rise of Crowdsourcing" as business practice:

> Technological advances in everything from product design software to digital video cameras are breaking down the cost barriers that once separated amateurs from professionals. Hobbyists, part-timers, and dabblers suddenly have a market for their efforts, as smart companies in industries as disparate as pharmaceuticals and television discover ways to tap the latent talent of the crowd. The labor isn't always free, but it costs a lot less than paying traditional employees. It's not outsourcing; it's crowdsourcing. (2006b)

Going on to write a book—*Crowdsourcing*—on the topic, Howe promulgates two definitions for the term on his website:

> **The White Paper Version:** Crowdsourcing is the act of taking a job traditionally performed by a designated agent (usually an employee) and outsourcing it to an undefined, generally large group of people in the form of an open call.

> **The Soundbyte Version:** The application of Open Source principles to fields outside of software.

These are radically different definitions, although Howe casually conflates them. In fact, many open source and open access advocates would likely claim that their principles are antithetical to Howe's "white paper version" of crowdsourcing. Framed as free labour, crowdsourcing can be seen as exploitative and, perhaps at its best, as an inefficient way of tackling academic workflows.[3] Crowdsourcing is often turned to by librarians and humanists as a way to facilitate bulk transcription or tagging—a sort of piecemeal work that can represent one problematic model of collaborative production. This is especially striking in the frequent identification of crowdsourcing as

---

[3] For an excellent blog series covering many of these issues, see Bentel (2014). For a perceptive analysis of the efficacy of crowdsourcing from an economic perspective, see Causer, Tonra, and Wallace (2012).

practice with Amazon's Mechanical Turk,[4] which pays workers very small amounts to complete small tasks. In disciplines more interested in human computation, algorithmic processing, or programming workflows, crowdsourcing is often taken to mean only the use of Mechanical Turk in algorithmic chains (Lasecki, Rzeszotarski, Marcus, and Bigham 2015; Law and Zhang 2011). Setting aside these debates about labour, however, is necessary because academic conceptions of crowdsourcing have far outgrown the corporate tool defined by Howe in 2006.

Broadly speaking, current usage of the term crowdsourcing has shifted; in academic contexts specifically, humanists often deploy the term with a remarkably different valence than Howe did more than a decade ago. At least as early as 2006, Howe himself began to notice that the term was being co-opted by others to mean broadly participatory types of digitally-facilitated interaction: crowdsourcing "is being used somewhat interchangably with Yochai Benkler's concept of commons-based peer production" (2006a). This separate term—commons-based peer production—was defined by Benkler and Nissenbaum as:

> a socio-economic system of production that is emerging in the digitally networked environment. Facilitated by the technical infrastructure of the Internet, the hallmark of this socio-technical system is collaboration among large groups of individuals, sometimes in the order of tens or even hundreds of thousands, who cooperate effectively to provide information, knowledge or cultural goods without relying on either market pricing or managerial hierarchies to coordinate their common enterprise. (2006, 394)

These are rather different cases of leveraging digital connectivity. Crowdsourcing, as originally understood, entailed a large business exploitatively drawing on "the crowd" to achieve a corporate-oriented purpose. Commons-based peer production suggests something different: namely, collaborative production of cultural or knowledge materials without distinct managerial or hierarchical oversight. This is perhaps easier to see when Benkler and Nissenbaum write that "peer production is a model of *social* production" (2006, 400; emphasis ours).

Now, crowdsourcing often encompasses commons-based peer production as well as more task-oriented conceptions of the term. This conflation can lead to misunderstandings about what is and is not possible within collaborative

---

[4] https://www.mturk.com/mturk/welcome.

environments, especially as straightforward, mechanical micro-tasks (e.g., categorizing, tagging, transcribing, geo-matching) scale up to more complex knowledge creation work (e.g., editing, creating controlled metadata, co-authoring). There are excellent examples of this kind of community-based peer production in the realm of digital humanities that work to blend research participation and undergraduate curriculum. The History Unfolded project, for example, seeks to have individuals retrieve local archival resources from across the U.S. in an effort to digitize and collect newspapers printed during WWII.[5] The potential to blend real world business interests with education are also present in the Planet Hunters website, which enables citizen scientists to identify planets on behalf of NASA's Kepler mission.[6] This is the creation of new knowledge, taking shape as the aggregate input of innumerable contributors. Social knowledge creation is a spectrum ranging from smaller, incremental contributions to fully shared intellectual ownership. Cultural production is a collaborative, networked practice, and when such interaction is made localized on a common platform or technical standard, the pathways of social knowledge creation are more easily traced. Central to any discussion of social media and social knowledge is an understanding of how profoundly normal such activity is now taken to be by the public.

These brief examples are also important because they point toward what Henry Jenkins and others have called "participatory culture." As outlined in a 2006 white paper, participatory culture is defined as "a culture with relatively low barriers to artistic expression and civic engagement, strong support for creating and sharing one's creations, and some type of informal mentorship whereby what is known by the most experienced is passed along to novices" (Jenkins, Purushotma, Clinton, Weigel, and Robison 2006, 3). More than existing as a type of subculture, life online is often participatory as a default; in other words, it is not a subset of cultural production, but the inclination of living a connected social and professional life online. History Unfolded and Planet Hunters are emblematic of the shift from individual ownership and an atomized field of production to an ethos of sharing and collaboration. These examples also signal a movement toward blending popular and scholarly culture that is, as many digital humanists have promised, more engaged and public facing. As experiments with crowdsourcing and the expectation to produce increasingly public scholarship illustrate, the humanities are attempting to catch up to where participatory culture

---

[5] https://newspapers.ushmm.org/.

[6] https://www.planethunters.org.

and commons-based production have already established substantial histories of practice.

Expanding Ideas of Collaboration

Faculty, students, administrators, parents, editors, publishers, and librarians might be considered broad classes and collections of individuals who impact scholarly communication. Documents in various genres and formats circulate through these communities: e-mails, PDFs, paper notes, dissertations, term papers, print-outs, books, journals, calendars, course schedules, syllabi, and so on. In addition to how these materials move, people move as well. They move to conference centres, to classrooms, to departmental offices, to the library, or across campus to the café. The formation of a scholarly community is complex and can appear opaque when scholars do not reflect publicly on their own practices. Of course, this is not to say that no work has been done. In a 1997 *PMLA* article, David R. Shumway wrote about the "star system in literary studies," noting that "[t]he emergence of academic stars, which has occurred only within the past twenty years, marks a fundamental shift in the profession of literary studies" (Shumway 1997, 90). His article ties together popular understandings of Hollywood stars, academic reputation, technological changes in jet travel, guidelines for tenure and promotion, and the rise of theory to explain what is often seen as simply "how things are." Shumway's intervention is valuable not for its insights into academic cultures per se, but because it is a mode of self-reflective analysis that provides a useful framing for our own discussions below. Material publications, actual people, specific events, and money are integral parts of academia and the knowledge it produces; Shumway's article is one way of approaching the formation and sustenance of academic behaviours and communities that are often uncommented on or else silently adopted.

In a similar reflexive vein, Elizabeth Losh has argued that "a document archive as a physical space is constituted by prohibitions on reading" (2004, 374). Losh focuses on the Bibliothèque nationale de France and the British Library to break down how the multiple levels of surveillance, access control, and physical design combine to form strictly controlled and unwelcome spaces. The main thrust of her argument is that such a gestalt cannot help but influence the development of digital libraries and archives; in other words, the politics of space and power bleed through to the creation of new media resources. Losh ends with a call for the type of research that is vital to understanding where the humanities has found itself, and where humanists of all stripes might go: "Perhaps more ethnographic approaches to the subject could be used as avenues to future research, in that our attitudes toward libraries, both real and virtual, as cultural institutions are shaped by shared beliefs about the function of physical spaces for public reading and by our epistemological expectations about how knowledge should be ordered" (2004, 384). Exploring the often tacit and unspoken ways in which scholars work, the materials they circulate, and the spaces they inhabit is crucial to consciously shaping the future of academic disciplines.

In an age of crowdsourcing, digital platforms, and constant connectivity, how do those working not only in the digital humanities, but the humanities more broadly, best remediate and re-present the cultural record? For digital humanists and scholarly editors, this is a pressing question. Remediation, representation, or modelling is an act of editing—as suggested by the fact that Wikipedia contributors are called editors. Scholarly editing has long been the domain of traditional scholars whose engagement with questions of media specificity, technology writ large, and disciplinary standards and knowledge construction has at times been underwhelming. The world of the print edition has reigned supreme in scholarly editing, and indeed in academia at large. But knowledge no longer moves only in the codex, and if we are to effectively bring the textual patrimony of our past into the present then we must understand how that is best accomplished. What's more, we need to envision the future shape of scholarly knowledge, broadly construed. What should become our guiding principles when we attempt to create knowledge-bearing objects such as the scholarly edition, or to design knowledge production systems? What can digital humanists, librarians, and #altac practitioners learn from initial forays into a scholarly landscape that may in the end operate according to radically different rules than the print-based ecosystem that has dominated for centuries? [7] Based on a number of trends in book history, scholarly editing, and digital humanities, this tendency toward shared ownership and participatory creation should be thought of as open scholarship. Less any final, distinct object than a set of interlocking processes, open scholarship—whether found in the form of articles, journals, monographs, new media projects, knowledge platforms, or data sets—is never finished, and far from controlled; instead, scholarship becomes a site of intersection for a number of activities which, taken together, indicate how

---

[7] "#alt-ac" describes those working on research projects with expert level training but are not occupying a typical faculty, library, or staff position within the university. These are knowledge workers that collect under a hash tag and are a symptom of larger trends in scholarly research labor in a digital age.

scholarly works might take shape in collaborative technological and social environments.

It has become a truism that the scholarly production and dissemination of knowledge is facing unprecedented economic difficulties. Among major scholarly organizations, the Modern Language Association (MLA) has recognized that the ground is shifting under formal disciplinary conventions in the humanities, and that the shift in scholarly publishing is being felt acutely by doctoral students just entering the field. The MLA Task Force on Doctoral Study in Modern Language and Literature met from 2012–2014, and was charged with exploring "the prospects for doctoral study in modern language and literature in the light of transformations in higher education and scholarly communication" (Modern Language Association 2014). Writing specifically about first academic books, but in a manner that is applicable to monograph publishing in general, Kathleen Fitzpatrick, Associate Executive Director and Director of Scholarly Communication at the MLA, writes: "The first academic book is [. . .] in a curious state, one that might usefully trouble our associations of obsolescence with the 'death' of this or that cultural form, for while the first academic book is no longer viable, it is still required. If anything, the first academic book isn't dead; it is undead" (2009). And if this is the case, Fitzpatrick wisely asks: "If the traditional model of academic publishing is not dead, but undead—again, not viable, but still required—how should we approach our work, and the publishing systems that bring it into being?" (2009.).

Although intellectual labour is often perceived by scholars to be unrelated to material production and financial exchange, it is anything but. The field of scholarly production and communication is one in which money and material circulation play a central role. It is the mismatch between internal, highly specific expectations concerning the form and type of scholarship, such as single author, peer-reviewed articles and monographs, and "external" factors such as material production and financial reality that Fitzpatrick addresses in *Planned Obsolescence*. In her persuasive view, "there is a particular form of book, the academic book—or more specifically (given that marketing departments prefer known quantities) the first academic book—that is indeed threatened with obsolescence" (Fitzpatrick 2009). In her discussions of new ways of facilitating peer review, understanding shared authorship practices, and a number of other topics, she often returns to the claim that what hinders progress in discovering new models of scholarly production is so often not technical but institutional.

We suggest that what the humanities faces is less a material obsolescence than an institutional one; we are caught in entrenched systems that no longer serve our needs. But because we are, by and large, our institutions, or rather, because they are us, the greatest challenge we face is not that obsolescence, but our response to it. Fitzpatrick's argument throughout *Planned Obsolescence* is vital because of how it reconceives authorship and textual production. Most relevant to this volume are her views on collaborative authorship, peer review, and the form that scholarly argumentation might take. In their briefest forms, she argues for a system of peer-to-peer review subtler than the current system; a reformulation of authorship that understands the collaborative nature of intellectual production; and a manner of networked knowledge production that fully leverages digital connectivity.

Social Knowledge in Humanities Practice

As much of the content in this volume describes, the way scholars in the humanities *do* scholarship is changing in response to the ways that those in the academy collectively leverage the fully-formed web infrastructure. Within the humanities, these trends can be most easily seen within the digital humanities, where social knowledge work seems most active. Lisa Spiro has written that "[b]uilding digital collections, creating software, devising new analytical methods, and authoring multimodal scholarship typically cannot be accomplished by a solo scholar; rather, digital humanities projects require contributions from people with content knowledge, technical skills, design skills, project management experience, metadata expertise, etc." (2009). She goes on to detail a roughly 45 per cent difference in collaborative authorship rates between *American Literary History* and *Literary and Linguistic Computing* (now titled *Digital Scholarship in the Humanities*). The first is a well-respected quarterly publication in literary studies; the second is the disciplinary journal for digital humanities. Although collaborative authorship is perhaps a rough proxy for discussing social knowledge creation more broadly, it is a useful snapshot of a community in practice over time. It is also representative of larger shifts of the sorts discussed by, among many others, Dan Cohen (2012), Kathleen Fitzpatrick (2009, 2012), Nancy Fjällbrant (1997), Jo Guldi (2013), and Ray Siemens (2002).

Digital humanists seem to have adopted the view that better connections between and among scholars, not to mention outside the academy, will and should irrevocably change the ways scholarship is pursued. There remains, however, a tendency to treat social media as a diversion, regardless of how the platform is used. Using online tools to ease collaboration and increase productivity has an additional side effect: by making analytics available to users, scholarly research production has become more transparent and quantifiable. It is now possible to use the user interface of social media in a secure and professional way through collaboration tools such as Slack, Basecamp, or Yammer. The Social Knowledge Timeline tool, created by a team at Penn State and led by Aaron Mauro, offers a way to link these common platforms with version control systems like Github and traditional social media like Twitter.[8] The roles and contributions of diverse individuals are much easier to trace in digital media than they have historically been in print. These common platforms allow for an accurate accounting of scholarly work, and produce a record of where insights emerge. Such traces are not new to the digital age, nor are they historically unique. Nevertheless, the evolution of print as a media form has often elided the contributions of multiple authors, textual producers, and the multiplicity of actors operating within and around print outputs. The sharpness of schematic critiques that combine reading and material production into an ecosystem of cultural work are notable in the humanities precisely because they explicitly trouble the flattening effect that print has on the contributions of collaborators, editors, designers, printers, and others.[9] Wiki-based publication platforms such as Wikibooks, or content management systems such as WordPress, help to reveal the patterns of social interactivity that have always typified knowledge work. Of course, as much as technology might reveal, technology also constitutes the relations it aids in uncovering. Culture and technology are mutually constitutive. Participating on a wiki makes the wiki work. Without human participation, the system is no more useful than traditional media. Because a wiki is designed to identify needed content, it grows in an organic way as users find areas of the wiki's knowledge base that is lacking. If an empty entry is found, the user is invited to begin the new entry. Participation is inspired by use. To state the obvious, knowledge creation occurs by knowing what is not known. Social knowledge creation helps build consensus about needed research activity by drawing questions, concerns, and discussion from a broader community in a rapid way. The speed of communication creates a qualitatively different experience for collaborators.

The speed of collaboration is often a thrilling experience for many humanists accustomed to working in isolation; however, it is also the traceability

---

[8] http://sktimeline.net/.

[9] Robert Darnton's communications circuit, for example, maps readers, authors, publishers, booksellers, printers, suppliers, and shippers onto a single diagrammatic visualisation of "the entire communication process" of books (1982, 67–68).

of scholarly work that reshapes social knowledge creation and the traditional hierarchies of universities through this reassessment of authority and
contributorship. It must be restated that all research is social. The broadly
understood roles of author, editor, and publisher are so often blended today, but when humanities scholars hold fast to strictly discursive concerns
and the role of the author, they fail to see the broader cultural impact that
technology is having in knowledge production and dissemination. This is not
to dismiss core disciplinary concerns. Humanities research must benefit in
some way by these new approaches, and these new approaches in no way
invalidate traditional methods. However, habituated methods run the risk
of isolating scholarly research from their contemporary technology and the
cultures that emerge in those contexts. It is impossible to encounter scholarship outside of a technological context, and every technological context
should be studied alongside and in relation to a sociocultural one.

Within humanities scholarship, and when referring to the technology produced and used by scholars themselves, exploring these types of relations
might appear contentious. A more subtle approach is to understand how
social technology manifests itself in local and particular terms, whether in
the print shop of early modern England or the academic department of the
contemporary university. Concerning digital scholarship, the social nature
of scholarly work is evident in the emphasis on collaboration. Many scholar-
practitioners in the field take such collaboration as a central tenet of the
discipline, and reports and reflections on the realities of digital humanities
projects and labs bolster such thinking.[10] Despite this emphasis on collaboration, as Spiro's observation highlights, explicitly social ways of working are
still seen as rare, on the whole, within the humanities.

Our challenge as practising humanists is to harmonize socially produced
intellectual work with the often rigid and outdated tenure and promotion
processes in our universities. Researchers of all ranks should be rewarded in
ways that are commensurate with their research output, regardless of their
chosen media or methods. Some have claimed that a media- and method-ag-
nostic culture will emerge as students resist outdated pedagogies and insist
on using new technologies in their education. Our students bring with them

---

[10] See chapter 3 ("The Social Life of the Digital Humanities") in Burdick et al. (2012,
73–98) for an excellent overview of the many ways "the social" manifests itself in
digital humanities. *Collaborative Research in the Humanities* (Deegan and McCarty 2012)
is exemplary, and indicative of disciplinary reflections on the nature of collaborative
work. See also Siemens's work on teams in digital humanities environments (2009).

deeply embedded patterns of behaviour and thinking that readily invalidate single authored scholarship. When social knowledge creation arrives ready-made and normalized, co-authorship and collaboration appear to be logical extensions of the technological context in which many scholars and students live and work every day. Humanists work to understand what it means to be human; in the twenty-first century, that means living connected and social lives online.

Knowledge work can be transparently social, outwardly iterative, and incredibly fast moving. Knowledge-making in the twenty-first century is shifting from long-understood modes of print production to collaborative, social systems of shared production and ownership. Matthew James Driscoll and Elena Pierazzo's recent collection *Digital Scholarly Editing*, for example, summarizes social editing in essentially two contexts: the first is the rise of social media connectivity and Web 2.0, and the second is developments in discourses of social textuality (2016, 18–25). The first set of ideas is obvious from the way many of us live our lives (there are nearly 1.5 billion monthly users of Facebook, for instance), and the second is perhaps best described by the work of Donald McKenzie (1999), Jerome McGann (1983, 2003, 2009), and Ray Siemens and collaborators (2002, 2009, 2010, 2011, 2014).

As a set of relationships in academia, social knowledge integrates both social media connectivity and the analytical categories scholars might bring to bear, in the manner of McKenzie or Robert Darnton, on the inscription-bearing artifacts of knowledge work in the humanities. In 2009, Jerome McGann repeated his earlier statement that "[i]n the next fifty years the entirety of our inherited archive of cultural works will have to be re-edited within a network of digital storage, access, and dissemination" (2009, 13). While the time scale may be ambitious, and seems to be more of a provocation than a plan of action, the realities of knowledge production indicate that such a transition is well underway. It must be noted, however, that McGann did not account for the role social knowledge creation has to play in this transition. To take an effective leadership role in this process of cultural remediation and reimagining, scholars must redefine humanities-based knowledge production as an expansive, welcoming process of collective cultural work. Instead of standalone models of individual authors, humanists must embrace a scholarship that is public facing, that integrates diverse groups into creative knowledge-making activities, and that is socially-minded.

An explicit goal of volumes 1 and 2 of *Social Knowledge Creation in the Humanities* is to lay bare the processes and decisions that form the cultures and communities surrounding digital humanities projects. By imagining "the social" as an analytical construct, humanists can take up a renewed popular role in remediating our "inherited archive of cultural works," and can reflect on internal practices and explore social knowledge as practice within the contemporary humanities. Approached in this way, social knowledge creation—whether it takes the form of editing, peer review, authorship, project building, or other activities within the ambit of digital humanities work—reimagines the making of knowledge in practical and institutional terms; social knowledge creation acknowledges the historical realities of knowledge work and takes account of contemporary trends in collaborative methods and social ownership of ideas. As an extension of scholarly communication, it cannot exist in the absence of communities of practice. In this regard, it is tied to material production, dissemination, and circulation of the history of ideas.

Digital humanists have a key role to play in designing and implementing platforms and projects that engender a shared public sphere of social knowledge creation. Participatory digital projects, and the collaborative or crowd-based infrastructures that underlie them, are mechanisms by which to re-knit and encourage shared ownership of cultural content. Researchers have a practical view of how knowledge is made and circulated. They see first-hand how disciplinary communities form and disintegrate, and how power is articulated in socially-oriented research groups.

Outline of Social Knowledge Creation in the Humanities Volumes

The two volumes of *Social Knowledge Creation in the Humanities* are being published in an iterative, unorthodox manner that combines printed outputs with digital publication. The final print form of all materials will be two printed volumes released by the New Technologies in Medieval and Renaissance Studies Series at Iter Press. Volume I of that set is the present document, which includes this introduction and a comprehensive bibliography of academic research on social knowledge creation broadly, and many of the areas touched on here (open access, crowdsourcing, coauthorship, etc.). Volume II will consist of a shorter introduction focused on the five contributions contained in that volume, as well as the chapters themselves.

We aim to simultaneously publish Volume 1 online and in print. In cooperation with Iter Press, the editors are building a website not only to disseminate these materials in digital form, but also to facilitate authorial and scholarly feedback using the well-established CommentPress theme in WordPress. Thus the online version will be open to commentary, feedback, and criticism.

The chapters intended for the printed Volume II will also be released at this time, on the same site. In publishing these materials online and on a platform that encourages open review and feedback, the authors, editors, and publishers hope that these materials will be enriched by such exchanges prior to revision for print publication. These exchanges will be integrated into the final print form, likely in the form of threaded marginal texts similar to glossing.

Due to the complexity of this process, it is worth outlining a full table of contents for the print volumes here:

Social Knowledge Creation in the Humanities

Volume I

Introduction: Tracing the Movement of Ideas: Social Knowledge Creation in the Humanities

An Annotated Bibliography of Social Knowledge Creation

Volume II

Introduction: Peer Reviewed and Expanded Proceedings of "Social Knowledge Creation in the Humanities" (a mini-conference sponsored by the Digital Humanities Summer Institute)

Future Radio and Social Knowledge Creation in the Humanities (John Barber)

Open Source Interpretation Using Z-axis Maps (Alex Christie)

Collocating Places and Words With TopoText (Randa El Khatib)

"Digital Zombies," A Learner-centred Game: Social Knowledge Creation at the Intersection of Digital Humanities and Digital Pedagogy (Juliette Levy)

The Page: Its Past and Future in Books of Knowledge (Christian Vandendorpe)

As such, here we introduce both printed volumes and the online site, which will facilitate extended conversation and exchange about social knowledge creation. The introduction to Volume II will address the contributions of that volume in more detail.

"An Annotated Bibliography of Social Knowledge Creation" is a foundational document in establishing the discursive parameters of an area of inquiry. Compiled by Alyssa Arbuckle, Nina Belojevic, Tracey El Hajj, Randa El Khatib, Lindsey Seatter, and Raymond G. Siemens, with Alex Christie, Matthew Hiebert, Jon Saklofske, Jentery Sayers, Derek Siemens, Shaun Wong, and the INKE and ETCL Research Groups, this bibliography is perhaps "best understood as a gestural environmental scan." It is, of necessity, "a partial snapshot" of scholarship that "continues to develop and coalesce around emerging areas of critical interest." And this scholarship is itself collaboratively created, an expanded iteration of previous work in the same area. Produced within a digital humanities research environment and relying on multiple digital tools to facilitate information sharing and academic writing, the bibliography annotates resources in three broad categories:

Social Knowledge Creation and Conveyance

Game-Design Models for Digital Social Knowledge Creation

Social Knowledge Creation Tools

In nearly 400 annotations, the authors provide "critical contexts and resources for the development of new tools, modes, and methods of scholarship that productively engage diverse communities." Many of the later chapters rely on the scholarship, platforms, and methodologies outlined and annotated in this bibliography. Published in print and online (as outlined above), the bibliography will become a resource consulted by academics, students, staff, and administration as social knowledge creation and open social scholarship evolve.

Conclusion: Technology as Practical Culture

In their postscript to the second edition of *Laboratory Life*, Bruno Latour and Stephen Woolgar write that the common assumption of succeeding laboratory studies such as theirs "is that our understanding of science can profitably draw upon experiences gained while immersed in the day-to-day activities of working scientists" (1979, 277). Latour and Woolgar self-consciously chose the term "anthropology of science" in an attempt to denote a particular "presentation of preliminary empirical material, our desire to retrieve something of the craft character of science, the necessity to bracket our familiarity with the object of study, and our desire to incorporate a degree of 'reflexivity' into our analysis" (1979, 277–78). To shift discussions of knowledge into discussions of social knowledge, the authors and resources in this volume have reflected on the spaces, technologies, and norms in daily use. Social knowledge production happens in social spaces (the lab, the centre), using social technologies (platforms of communication, research, discovery, analysis, etc.), and employs variants on the expectations and norms arising from specific disciplines. Central to discussions of the changing nature of scholarly communication and knowledge creation is how the infrastructures for making and sharing knowledge in the humanities are shifting.[11] The transitions seen in academic libraries are an important early symptom of this largely uncharted paradigm. Overall, the academy needs to better understand the internal systems of validating digital scholarship. Further, there is a necessity for community-based consensus around structuring student work and collaboration between teams, and large projects need to state the processes by which knowledge is made and shared. There may be resistance to this type of introspective analysis, in part because the humanities have always been accused of navel gazing, and also because our methods will demand a great deal of change from university systems around the world. The challenge of institutional transformation is neatly articulated by Geoffrey C. Bowker and Susan Leigh Star in their study of classification:

> Remarkably for such a central part of our lives, we stand for the most part in formal ignorance of the social and moral order created by these invisible, potent entities. Their impact is indisputable, and as Foucault reminds us, inescapable. Try the simple experiment of ignoring your gender classification and use instead whichever toilets are the nearest; try to locate a library book shelved under the wrong Library of Congress catalogue number; stand in the immigration queue at a busy foreign airport without the right passport or arrive without the transformer and the adaptor that translates between electrical standards. The material force of categories appears always and instantly. (1999, 3)

---

[11] Initiatives such as the Canadian Research Knowledge Network's (CRKN) Integrated Digital Scholarly Ecosystem (IDSE) project demonstrate the comprehensive difficulties of such a shift; the role of libraries, for example, seems to be shifting from that of a body that houses content and provides access to one that aids in co-creating and fostering scholarship. This is one example of how knowledge work happens in a specific, but often tacit, system of standards, expectations, organization, and spatial arrangement.

This discussion of social knowledge creation is an effort to reveal some of the assumptions that underwrite humanities research by illustrating its emergence in locally specific ways. We embrace, in this volume, an emerging social order in humanities research. We acknowledge and validate the technology and methods that form and inform our academic cultures. We follow in the footsteps of other interdisciplinary scholars. The words of Ursula Franklin, first heard in her Massey Lectures of 1989, resonate throughout the material that follows:

> Looking at technology as practice, indeed as formalized practice, has some quite interesting consequences. One is that it links technology directly to culture, because culture, after all, is a set of socially accepted practices and values. Well laid down and agreed upon practices also define the practitioners as a group of people who have something in common because of the way they are doing things. (2004, 6)

This volume, much like Lovejoy's *The Great Chain of Being*, represents a foundational discursive moment. But it is a discursive moment that breaks the chain and embraces the organic and non-hierarchical networks that emerge between people. By articulating a definition of social knowledge creation that takes shape at the intersection of multiple trends and scholarly practices, the work collected here is in some sense creating a new area for discussion in digital humanities. Our goal is not to have the final word on this debate, but to provide initial, gestural, and varied snapshots of scholarly practices in flux.

WORKS CITED

Barthes, Roland. 1977. *Image, Music, Text*. Translated by Stephen Heath. London: Fontana Press.

Benkler, Yochai, and Helen Nissenbaum. 2006. "Commons-based Peer Production and Virtue." *Journal of Political Philosophy* 14 (4): 394–419.

Bentel, Nikolas. 2014. "The Problem of Crowdsourcing, Part 1: A Modern Form of Exploitation." *HASTAC*. https://www.hastac.org/blogs/nikolas-bentel/2014/04/01/problem-crowdsourcing-part-1-modern-form-exploitation.

Bosch, Stephen, and Kittie Henderson. 2015. "Whole Lotta Shakin' Goin' On: Periodicals Price Survey 2015." http://lj.libraryjournal.com/2015/04/publishing/whole-lotta-shakin-goin-on-periodicals-price-survey-2015/.

Bowker, Geoffrey C., and Susan Leigh Star. 1999. *Sorting Things Out: Classification and its Consequences.* Cambridge, MA: MIT Press.

Burdick, Anne, Johanna Drucker, Peter Lunenfeld, Todd Presner, and Jeffrey Schnapp. 2012. "The Social Life of the Digital Humanities." In *Digital_Humanities,* 73–98. Cambridge, MA: MIT Press.

Causer, Tim, Justin Tonra, and Valerie Wallace. 2012. "Transcription Maximized; Expense Minimized?: Crowdsourcing and Editing *The Collected Works of Jeremy Bentham.*" Digital Scholarship in the Humanities (formerly *Literary and Linguistic Computing*) 27 (2): 119–37.

Cohen, Daniel J. 2010. "Open Access Publishing and Scholarly Values" [blog post]. May 27. http://www.dancohen.org/2010/05/27/open-access-publishing-and-scholarly-values/.

_____ 2012. "The Social Contract of Scholarly Publishing." In *Debates in the Digital Humanities,* edited by Matthew K. Gold, 319–21. Minneapolis: University of Minnesota Press. http://dhdebates.gc.cuny.edu/debates/text/27.

Crow, Raym. 2009. "Income Models for Open Access: An Overview of Current Practice." Scholarly Publishing and Academic Resources Coalition (SPARC). http://www.sparc.arl.org/sites/default/files/incomemodels_v1.pdf.

Darnton, Robert. 1982. "What is the History of Books?" *Daedalus* 111 (3): 65–83. http://www.jstor.org/stable/20024803.

Deegan, Marilyn, and Willard McCarty, eds. 2012. *Collaborative Research in the Digital Humanities.* Farnham, UK, and Burlington, VT: Ashgate.

Driscoll, Matthew James, and Elena Pierazzo, eds. 2016. *Digital Scholarly Editing: Theories and Practices.* Cambridge: Open Book Publishers.

Fitzpatrick, Kathleen. 2009. *Planned Obsolescence: Publishing, Technology, and the Future of the Academy.* http://mcpress.media-commons.org/plannedobsolescence/introduction/undead/.

_____ 2012. "Beyond Metrics: Community Authorization and Open Peer Review." In *Debates in the Digital Humanities,* edited by Matthew K. Gold, 452–59. Minneapolis: University of Minnesota Press. http://dhdebates.gc.cuny.edu/debates/text/7.

Fjällbrant, Nancy. 1997. "Scholarly Communication—Historical Development and New Possibilities." In *Proceedings of the IATUL Conferences.* West Lafayette, IN: Purdue University Libraries e-Pubs. http://docs.lib.purdue.edu/cgi/viewcontent.cgi?article=1389&context=iatul.

Franklin, Ursula M. *The Real World of Technology.* (1990) 2004. Revised edition. Toronto: House of Anansi Press.

Guldi, Jo. 2013. "Reinventing the Academic Journal." In *Hacking the Academy: New Approaches to Scholarship and Teaching from Digital Humanities,* edited by Daniel J. Cohen and Tom Scheinfeldt, 19–24. Ann Arbor: University of Michigan Press. http://dx.doi.org/10.3998/dh.12172434.0001.001.

Howe, Jeff. 2006a. "Crowdsourcing: A Definition" [blog post]. *Crowdsourcing: Why the Power of the Crowd is Driving the Future of Business.* http://crowdsourcing.typepad.com/cs/2006/06/crowdsourcing_a.html.

_____ 2006b. "The Rise of Crowdsourcing." *Wired.* June 1. http://archive.wired.com/wired/archive/14.06/crowds.html.

_____ n.d. *Crowdsourcing: Why the Power of the Crowd is Driving the Future of Business* [blog]. http://crowdsourcing.com/.

INKE [Implementing New Knowledge Environments]. 2017. "Victoria Gathering 2017." http://inke.ca/projects/victoria-gathering-2017/.

Jenkins, Henry. 2006. *Convergence Culture.* New York: New York University Press.

Jenkins, Henry, with Ravi Purushotma, Katherine Clinton, Margaret Weigel, and Alice J. Robison. 2006. "Confronting the Challenges of Participatory Culture: Media Education for the 21st Century." Chicago: MacArthur Foundation. http://www.newmedialiteracies.org/wp-content/uploads/pdfs/NMLWhitePaper.pdf.

Kennison, Rebecca, and Lisa Norberg. 2014. "A Scalable and Sustainable Approach to Open Access Publishing and Archiving for Humanities and Social Sciences: A White Paper." April 11. New York: K|N Consultants.

http://knconsultants.org/wp-content/uploads/2014/01/OA_
Proposal_White_Paper_Final.pdf.

Lasecki, Walter S., Jeffrey M. Rzeszotarski, Adam Marcus, and Jeffrey P.
Bigham. 2015. "The Effects of Sequence and Delay on Crowd Work."
In *Proceedings of the 33rd Annual ACM Conference on Human Factors in
Computing Systems (CHI '15)*, 1375–78. New York: ACM.

Latour, Bruno, and Stephen Woolgar. 1979. *Laboratory Life: The Construction of
Scientific Facts.* Beverly Hills, CA: Sage Publications.

Law, Edith, and Haoqi Zhang. 2011. "Towards Large-Scale Collaborative
Planning: Answering High-level Search Queries Using Human
Computation." In *Proceedings of the Twenty-Fifth AAAI Conference on
Artificial Intelligence (AAAI '11)*, 1210–15. Palo Alto, CA: AAAI Publications.

Losh, Elizabeth. 2004. "Reading Room(s): Building a National Archive in
Virtual Spaces and Physical Places." *Digital Scholarship in the Humanities*
(formerly *Literary and Linguistic Computing*) 19 (3): 373–84.

Lovejoy, Arthur O. 1936. *The Great Chain of Being: A Study of the History of an
Idea.* Cambridge, MA: Harvard University Press.

Mahony, Nick, and Hilde C. Stephansen. 2016. "The Frontiers of Participatory
Public Engagement." *European Journal of Cultural Studies.* n.p.
doi:10.1177/1367549416632007.

Maxwell, John W. 2015. "Beyond Open Access to Open Publication and Open
Scholarship." *Scholarly and Research Communication* 6 (3): n.p. http://
src-online.ca/index.php/src/article/view/202.

McGann, Jerome J. 1983. *A Critique of Modern Textual Criticism.* Chicago: Chicago
University Press.

_____ 2003. "Textonics: Literary and Cultural Studies in a Quantum
World." In *The Culture of Collected Editions*, edited by Andrew Nash, 245–
60. Basingstoke, UK: Palgrave Macmillan.

_____ 2009. "Our Textual History." *Times Literary Supplement* 5564:
13–15.

McKenzie, D. F. 1999. *Bibliography and the Sociology of Texts.* Cambridge:
Cambridge University Press.

Meadows, Alice. 2015. "Beyond Open: Expanding Access to Scholarly Content." *Journal of Electronic Publishing* 18 (3): n.p. http://quod.lib.umich.edu/j/jep/3336451.0018.301?view=text;rgn=main.

Modern Language Association (MLA). 2014. *Report of the Task Force on Doctoral Study in Modern Language and Literature*. New York: MLA. https://www.mla.org/Resources/Research/Surveys-Reports-and-Other-Documents/Staffing-Salaries-and-Other-Professional-Issues/Report-of-the-Task-Force-on-Doctoral-Study-in-Modern-Language-and-Literature-2014.

Organisation for Economic Co-operation and Development (OECD). 2015. "Making Open Science a Reality." *OECD Science, Technology and Industry Policy Papers* 25: n.p. doi:10.1787/23074957.

Research Councils UK. n.d. "Pathways to Impact" [implemented in 2009]. http://www.rcuk.ac.uk/innovation/impacts.

Saklofske, Jon, and the INKE Research Team. 2016. "Digital *Theoria, Poiesis,* and *Praxis*: Activating Humanities Research and Communication through Open Social Scholarship Platform Design." *Scholarly and Research Communication* 7 (2/3): n.p. http://src-online.ca/index.php/src/article/view/252/495.

Scott, John. 1991. *Social Network Analysis: A Handbook.* London and Beverly Hills, CA: Sage Publications.

Shumway, David R. 1997. "The Star System in Literary Studies." *PMLA* 112 (1): 85–100.

Siemens, Lynne. 2009. "It's a Team if You Use 'Reply All': An Exploration of Research Teams in Digital Humanities Environments." *Digital Scholarship in the Humanities* (formerly *Literary and Linguistic Computing*) 24 (2): 225–33. doi:10.1093/llc/fqp009.

Siemens, Raymond G. 2002. "Scholarly Publishing at its Source, and at Present." In *The Credibility of Electronic Publishing: A Report to the Humanities and Social Sciences Federation of Canada*, compiled by Raymond G. Siemens, Michael Best, Elizabeth Grove-White, Alan Burk, James Kerr, Andy Pope, Jean-Claude Guédon, Geoffrey Rockwell, and Lynne Siemens. *TEXT Technology* 11 (1): 1–128.

Siemens, Raymond G., Mike Elkink, Alastair McColl, Karin Armstrong, James Dixon, Angelsea Saby, Brett D. Hirsch and Cara Leitch, with Martin Holmes, Eric Haswell, Chris Gaudet, Paul Girn, Michael Joyce, Rachel Gold, and Gerry Watson, and members of the PKP, Iter, TAPoR, and INKE teams. 2010. "Underpinnings of the Social Edition? A Narrative, 2004–9, for the Renaissance English Knowledgebase (REKn) and Professional Reading Environment (PReE) Projects." In *Online Humanities Scholarship: The Shape of Things to Come*, edited by Jerome J. McGann, with Andrew Stauffer, Dana Wheeles, and Michael Pickard. Houston: Rice University Press. http://cnx.org/content/m34335.

_____ 2011. "Prototyping the Renaissance English Knowledgebase (REKn) and Professional Reading Environment (PReE), Past, Present, and Future Concerns: A Digital Humanities Project Narrative." *Digital Studies / Le champ numérique* 2 (2): n.p. http://www.digitalstudies.org/ojs/index.php/digital_studies/article/view/182/255.

_____ 2014. "Underpinnings of the Social Edition? A *Brief* Narrative, 2004–9, for the Renaissance English Knowledgebase (REKn) and Professional Reading Environment (PReE) Projects, and a Framework for Next Steps." In *New Technologies and Renaissance Studies II*, edited by Tassie Gniady, Kris McAbee, and Jessica Murphy, 3–49. *New Technologies in Medieval and Renaissance Studies* 4. Toronto: Iter Academic Press; Tempe: Arizona Center for Medieval and Renaissance Studies.

Siemens, Raymond G., Johanne Paquette, Karin Armstrong, Cara Leitch, Brett D. Hirsch, Eric Haswell, and Greg Newton. 2009. "Drawing Networks in the Devonshire Manuscript (BL Add 17492): Toward Visualizing a Writing Community's Shared Apprenticeship, Social Valuation, and Self-Validation." *Digital Studies / Le champ numérique* 1 (1): n.p. http://www.digitalstudies.org/ojs/index.php/digital_studies/article/view/146/201.

Spiro, Lisa. 2009. "Collaborative Authorship in the Humanities" [blog post]. *Digital Scholarship in the Humanities*, April 21. http://digitalscholarship.wordpress.com/2009/04/21/collaborative-authorship-in-the-humanities.

UK Research Excellence Framework. 2012. "Assessment Framework and Guidance on Submissions." [Published in July 2011; updated in January 2012.] http://www.ref.ac.uk/media/ref/content/

pub/assessmentframeworkandguidanceonsubmissions/GOS%20 including%20addendum.pdf.

Willinsky, John. 2006. *The Access Principle: The Case for Open Access to Research and Scholarship.* Cambridge, MA, and London: MIT Press.

An Annotated Bibliography of
Social Knowledge Creation

Alyssa Arbuckle, Nina Belojevic, Tracey El Hajj, Randa El
Khatib, Lindsey Seatter, and Raymond G. Siemens, with Alex
Christie, Matthew Hiebert, Jon Saklofske, Jentery Sayers, Derek
Siemens, Shaun Wong, and the INKE and ETCL Research Groups

Introduction

In 2015–16, a collaborative team at the Electronic Textual Cultures Lab (ETCL)[1] endeavoured to update the previously published "Social Knowledge Creation: Three Annotated Bibliographies."[2] The former tripartite annotated bibliography collection was developed in 2012–13 at the ETCL, in collaboration with the Implementing New Knowledge Environments (INKE) research group.[3]

---

[1] The Electronic Textual Cultures Lab (ETCL) is located at the University of Victoria in Victoria, BC, Canada. The ETCL engages in cross-disciplinary study of the past, present, and future of textual communication, and is a hub for digital humanities activities across the University of Victoria campus and beyond. It is a digital humanities lab with research, teaching, and service mandates, and an intellectual centre for the activities of some 20 local faculty, staff, students, and visiting scholars (more than 60 since inception), who work closely with research centres, libraries, academic departments, and projects locally and in the larger community. Through a series of highly collaborative relationships, the ETCL's international research community comprises more than 300 researchers. Dr. Raymond G. Siemens directs the ETCL.

[2] Alyssa Arbuckle, Nina Belojevic, Matthew Hiebert, and Raymond G. Siemens, with Shaun Wong, Derek Siemens, Alex Christie, Jon Saklofske, Jentery Sayers, and the INKE and ETCL Research Groups, "Social Knowledge Creation: Three Annotated Bibliographies," *Scholarly and Research Communication* 5, no. 2 (2014), http://src-online.ca/index.php/src/article/view/150/299.

[3] Implementing New Knowledge Environments (INKE; inke.ca) is a collaborative group of researchers and graduate research assistants working with other organizations and partners to explore the digital humanities, electronic scholarly communication, and the affordances of digital text. INKE is directed by Dr. Raymond G. Siemens and funded by a seven-year Social Sciences and Humanities Research Council (SSHRC) Major Collaborative Research Initiatives grant.

ISBN 978-0-86698-739-4 (online) ISBN 978-0-86698-583-3 (print)
New Technologies in Medieval and Renaissance Studies 7 (2017) 29–264

Although "Social Knowledge Creation: Three Annotated Bibliographies" provided a thorough snapshot of scholarship and initiatives related to social knowledge creation up to the year 2013, this area of inquiry has expanded steadily since that time, and requires further attention. Now revisited and renewed, "An Annotated Bibliography of Social Knowledge Creation" aims to broaden the conceptual scope of social knowledge creation, and makes notable additions and expansions—including those in the subjects of public humanities, crowdsourcing, collaborative games, digital publishing, and open access. Throughout this process, we have considered social knowledge creation as follows: acts of collaboration in order to engage in or produce shared cultural data and/or knowledge products. The authors enacted social knowledge creation practices in the very development of the current document. Not only did we engage, reshape, and build on the previous authors' work, but we also collaborated on the intellectual direction of this iteration, and the compilation of new resources. This work was facilitated by electronic authoring platforms such as Google Drive,[4] and integrated the substantial Zotero[5] bibliographical library developed by the previous authorial team, which served as a critical foundation for this revised publication.

The previous document, "Social Knowledge Creation: Three Annotated Bibliographies," gathered and annotated bibliographical items as a resource for students and researchers interested in INKE research areas, including participants at digital humanities seminars in Bern, Switzerland (June 2013), and Leipzig, Germany (July 2013).[6] Updated as "An Annotated Bibliography of Social Knowledge Creation," the result of these initiatives might be best understood as a gestural environmental scan. Necessarily a partial snapshot, the scholarship reflected continues to develop and coalesce around emerging areas of critical interest. Neither the current nor the previous iteration claims to establish a canon; rather, we hope to provide a glimpse into interconnected concepts and fields through collaborative aggregation and annotation. Given the nature of social knowledge creation as a research area, many of the bibliographical entries span several of our delineated categories. In those cases, the entries have been duplicated and appear in all of their

---

[4] https://www.google.com/drive/.

[5] https://www.zotero.org/.

[6] Of note, the game-focused section of "Social Knowledge Creation: Three Annotated Bibliographies" was also revised for the journal *Mémoires du livre / Studies in Book Culture*, and published as "A Select Annotated Bibliography Concerning Game-Design Models for Digital Social Knowledge Creation" (Belojevic, Arbuckle, Hiebert, Siemens, et al., 2014).

respective categories to ensure easy access, readability, and the most complete treatment possible of each category. Please note that duplicated entries are marked with an asterisk (*) after their first appearance.

Intent

We revised and updated "An Annotated Bibliography of Social Knowledge Creation" in order to offer critical contexts and resources for the development of new tools, modes, and methods of scholarship that engage diverse communities productively, including researchers, students, librarians, faculty, administrators, academically-aligned and public groups, and engaged members of the public. Our exploration was guided by the following questions: How can scholarly activity in online environments be modelled for greater public engagement? What can we learn from past and existing knowledge creation practices? Will the humanities continue to play a leading role in knowledge production within transforming social landscapes? How can we theorize the ongoing changes in knowledge conditions in ways that might account for our critical design-based interventions? What existing humanities processes should our new knowledge environments seek to redesign? How do social knowledge creation practices perpetuate an inclusive environment, and how do they inadvertently exclude marginalized groups?

We also commit to certain overarching concepts: knowledge is plural and its outputs are variegated, as emphasized by much of the research included (e.g., Gitelman 2006; McKenzie 1999). Rather than conceptualizing knowledge as singular ("the Media"), we consider knowledge as a multifaceted product of many social, material, and media forces. To take one example of many, there was no single "print culture" animating the world inaugurated by Gutenberg, but rather a myriad of localized print cultures (Johns 1998)—a claim Adrian Johns makes in firm opposition to the more straightforward cause-and-effect theory put forth by Elizabeth Eisenstein in *The Printing Press as an Agent of Change* (1979). In considering these lines of inquiry, we learned quickly that social knowledge creation does not necessarily heed traditional disciplinary boundaries and practices; rather, it cuts across research areas and methods, and emphasizes the innate value of collaboration.

Section 1: Social Knowledge Creation and Conveyance

Popular culture and social activity are often located in digital spaces, rendering knowledge creation increasingly collaborative and plural. As such, multiple institutions, cultural specificities, and political and economic conditions affect knowledge production in idiosyncratic ways (Burke 2000). In the twenty-first century, much of culture is developed, formalized, and perpetuated through the global network of the Internet as the primary space of knowledge creation, engagement, and dissemination. The prevalence of dynamic social knowledge creation and digital tools provides new opportunities for the humanities and social sciences to connect with multiple publics, who might not necessarily identify as academic.

In the "Social Knowledge Creation and Conveyance" section of the bibliography, we gather perspectives on how knowledge creation processes might be imagined for digital environments. We also consider older forms of communication and interaction from the history of knowledge production. For instance, conversation, epistolary correspondence, manuscript circulation, and other informal modes of scholarly exchange have been recovered at the fount of academic disciplines (Siemens 2002). Various similarities can be drawn between manuscript culture and digital knowledge creation, and many electronic publications can enable social knowledge creation through critical engagement with the social elements and lessons from the history of scholarly communication. Both practices are founded on a cyclical process of drafting, editing, and revising in which documents are revisited continually, rather than a print model that perpetuates the idea of an "end product" (i.e., the published work). This writing practice is exemplified in the work of Jane Austen, who returned to her juvenile writings throughout her lifetime (Levy 2010), as well as in the eighteen surviving, distinct versions of Samuel Taylor Coleridge's "The Rime of the Ancient Mariner" that were drafted and circulated between November 1797 and March 1798 (Stillinger 1994). Moreover, both manuscript culture and web technologies rely explicitly on the social characteristics of knowledge creation, as authors seek the critical input of other experts to shape their ideas. Today, collaborative drafting and editing platforms such as Google Drive marry these conventions by providing an environment in which we are able to write, revise, and comment on our own work and the work of others simultaneously—a literary circle of the twenty-first century variety. As we witness digital transformations, the principles of older technologies and cultural practices emerge and manifest in reimagined ways.

How might we model these practices in the humanities? To explore this question, we have included resources that focus on how users may be empowered by social knowledge creation tools in order to contribute to knowledge production in online environments. Methodological practices of scholarship in all disciplines are increasingly affected by common digital affordances (McCarty 2005). Current web technologies offer a degree of interaction, collaboration, and sociality that contrasts with the early static websites modelled on the written page, which often privilege straightforward and linear information conveyance. The trend toward greater access to data in widely usable formats, and the growing familiarity with analytical tools to process that data, dramatically accelerates workflows and allows researchers to pose questions that simply would have taken too long to answer without computation. Online, interactive publication models are also gaining popularity for those who wish to actively engage the larger public in the sharing of academic work, and who may feel as though corporate academic publishers' stakes in traditional journals and monographs have dominated scholarly knowledge production negatively. The increasing use by researchers of software-based modes for communication can be seen to cultivate a "problem-based" approach to scholarship that locates focus and concern outside of disciplinary boundaries. Problem-based scholarship implies greater attunement with the public that research intends to serve, and in so doing suggests that deepening discourse between experts and the communities that exist around data sets is valuable.

The ETCL and INKE teams have explored the study and practice of social knowledge creation through their public development of *A Social Edition of the Devonshire Manuscript* on Wikibooks (Bowen, Crompton, and Hiebert, 2014; Crompton, Siemens, Arbuckle, et al. 2015; Crompton, Siemens, Arbuckle, et al. 2013; Siemens, Timney, and Leitch, et al. 2012; see "The Shifting Future of Scholarly Communication and Digital Scholarship" and "Social Knowledge Creation in Electronic Scholarly Editions and e-Books").[7] By prototyping an edition of an early modern text on the principles of open access and editorial transparency, this model suggests that new media environments can effectively facilitate access, contribution, and discussion of scholarly knowledge for stakeholders both within and beyond the academy without jettisoning quality or the peer review process. The social edition engages discourse surrounding scholarly knowledge production, new media, and critical making to develop an argument about the nature of scholarly editing. In extending the dynamic relations inherent to textual production and reception, the social scholarly edition transforms the role of the editor from that of a didactic authority to that of a knowledge creation facilitator. In other examples, the *Transcribe Bentham* project unites a humanities research aim with public engagement by developing a digital platform that facilitates the crowdsourced transcription of political philosopher Jeremy Bentham's manuscripts (see the "Crowdsourcing" category); the *OpenStreetMap* project brings together

---

[7] See https://en.wikibooks.org/wiki/The_Devonshire_Manuscript.

crowdsourcing and digital mapping in its collection of over two million user uploads of geospatial data (see the "Spatial Humanities and Digital Mapping" category). Situating scholarly knowledge creation practices in a public, social sphere involves a redistribution of authority that gestures toward universal inclusiveness, and reactivates open, community-based collaborative processes by means of digital tools. In their digitally networked forms, basic scholarly activities—"scholarly primitives" as John Unsworth (2000) has termed them—extend into the public sphere. This shift in praxis unsettles conceptions of the researcher as a sovereign discoverer of knowledge in an objective world.

Social knowledge creation activities embody the epistemological and institutional changes occurring within society, but hinge on stakeholders building capacity in order to use and develop new methodologies and forms of communication. The involvement of citizen scholars is crucial, especially if humanities scholars consider social knowledge creation as an opportunity to reinvigorate and sustain research and dissemination. In this way, social knowledge creation can become an effective response to the perennial call for higher education institutions to actively engage community groups. In contrast to the undermining effects of corporate-based funding, economic incentives, or the commodification of the humanities, we argue that public involvement in knowledge creation can productively bolster the humanities through integration with its traditional values (Brown 1995; Ellison 2008; Farland 1996).

Section 2: Game-Design Models for Digital Social Knowledge Creation

Another key area of concern for social knowledge creation is borne among earlier notions that the book is, itself, an inherently social technology. This concept is embodied in nascent efforts to digitally render the collaborative nature of textual analysis in projects such as *Ivanhoe*, an online environment for community-based literary analysis created by Johanna Drucker and Jerome McGann (Drucker 2003). The *Ivanhoe* project was the first popular attempt at this kind of socially gamified reading. Its very nature points to the potential for game-based design techniques to more broadly assist in modelling collaborative scholarly interpretation practices. Following this, the second section of the bibliography, "Game-Design Models for Digital Social Knowledge Creation," incorporates critical assessment of the role of gaming in social knowledge creation, as it is essential for moving forward in the scholarly development of game-design models for publication and communication. We seek game-design techniques that might effectively contribute to humanities-based knowledge practices, as well as consider how gaming as a cultural phenomenon is capable of constituting subjects in ways that might perpetuate exploitative labour dynamics and rigidified knowledge regimes.

Game elements such as badges and achievements have inspired alternative recognition systems within non-game scholarly contexts to increase participation. This is exemplified in the growing use of altmetrics—a movement rooted in scientific journal practices—for quantifying article impact data. A number of critics from within the humanities have condemned such use of gamification for corroding the intrinsic motivation of knowledge-building activities. Conversely, other scholars argue that the processes of gamification attenuate the inherent power of full games to convey knowledge, make arguments, and accomplish other meaningful objectives (Bogost 2011). Theorists wishing to retain gamification as a sociological or media theory concept—to account, for instance, for the unique experiential phenomenon of "flickering" between game and non-game contexts (Deterding, Dixon, Khalad, and Nacke 2011)—have developed terminology distinct from "gamification," and aim to limit its range of applicable techniques to the use of non-achievement-related game elements within scholarly knowledge environments. How we analyze and understand past and present knowledge environments may be reconstituted through game-design and implementation, thus fostering the dialectical relationship between the critical and creative aspects of social knowledge production in digital environments.

Section 3: Social Knowledge Creation Tools

Social knowledge creation is often provoked through collaborative electronic tools that model such processes as John Unsworth's "scholarly primitives": discovering, annotating, comparing, referring, sampling, illustrating, and representing (2000). Creation-as-research and critical making are emerging as key practices of social knowledge creation (Chapman and Sawchuk 2015; Ratto, Wylie, and Jalbert 2014), and can be embedded as computational tools. More and more frequently, we see the integration of developing trends and technologies in scholarly exploration, such as the popularity of 3D printing in libraries and labs. Critical making is recognized for its popular draw and positive social impact, and will most likely continue to grow and provoke change in different arenas (Ratto and Ree 2012). The tools highlighted in the final section of the annotated bibliography, "Social Knowledge Creation Tools," encourage the construction of flexible, open systems that evolve alongside the knowledge creation they facilitate. An advantage of using specifically open source tools in this context is that they are fundamentally participatory, allow arguments to be transparent at the level of code, and include adjustable, adaptable process modelling.

The open source community revolves around several collaborative code repositories used for developing and distributing software, such as Source-Forge and GitHub. As Alan Galey and Stan Ruecker (2010) argue, the *design* of prototyped digital tools perpetuates certain politics or makes specific arguments about the processes it intends to model. The social knowledge creation tools included in this section derive from areas of content provision, annotation, marking/tagging, bibliography, and text analysis. Collaborative tools for annotation model a scholarly primitive that emerged with medieval manuscript culture to assist in remembering, thinking, clarifying, sharing, and interpretating (Ovsiannikov, Arbib, and McNeill 1999; Marshall 1997; Wolfe 2002). Blogs and content management systems facilitate user-derived content, contending that sharing, creativity, and dialogue are intrinsic to knowledge work (Fitzpatrick 2007; Kjellberg 2010; Fernheimer, Litterio, and Hendler 2011). Collaborative bibliography tools enhance the scholarly processes they model by heightening social involvement and reflecting the networked nature of thought and scholarship (Cohen 2008; Hendry, Jenkins, and McCarthy 2006). Community bibliography applications, which often incorporate folksonomy tagging, allow for the collaborative creation, organization, citation, enrichment, and publication of bibliographies. Applied social knowledge creation tools for textual analysis involve "the application of algorithmically facilitated search, retrieval, and critical processes" (Schreibman, Siemens, and Unsworth 2008). After the profound "changes in knowledge regimes" of recent decades (Burke 2000), situated users are increasingly capable of redefining what media become, regardless of the publics they are initially constructed for and aim to construct (Gitelman 2006).

Conclusion

Facilitating public involvement in scholarship through digital means can encourage the humanities to ask the "right" questions, provide better means of answering them, and improve competency in reflecting such answers in both expert communities and larger societal discourses. New media facilitates interactive public humanities practices with greater ease, as within the digital realm "the presentation of knowledge and the production of knowledge happen interdependently and simultaneously" (Jay 2012). This implied transformation of scholarship invites renewed inquiry into the field of knowledge production. Such inquiry, we believe, might inform practitioners in their efforts to create critical interventions (including through content modelling, critical processes, and communication and dissemination) that best facilitate a convergence of the scholarly and the social spheres while preserving a commitment to humanities-based research.

WORKS CITED

Belojevic, Nina, Alyssa Arbuckle, Matthew Hiebert, Raymond G. Siemens, Shaun Wong, Alex Christie, Jon Saklofske, Jentery Sayers, and Derek Siemens, with the INKE and ETCL Research Groups. 2014. "A Select Annotated Bibliography Concerning Game-Design Models for Digital Social Knowledge Creation," *Mémoires du livre/Studies in Book Culture* 5 (2): n.p. http://www.erudit.org/revue/memoires/2014/v5/n2/1024783ar. html?vue=plan.

Bogost, Ian. 2011. *How to Do Things with Videogames.* Minneapolis: University of Minnesota Press.

Bowen, William R., Constance Crompton, and Matthew Hiebert. 2014. "Iter Community: Prototyping an Environment for Social Knowledge Creation and Communication." *Scholarly and Research Communication* 5 (4): n.p. http://src-online.ca/index.php/src/article/view/193/360.

Brown, David W. 1995. "The Public/Academic Disconnect." In *Higher Education Exchange Annual,* 38–42. Dayton, OH: Kettering Foundation.

Burke, Peter. 2000. *A Social History of Knowledge: From Gutenberg to Diderot.* Cambridge: Polity Press.

Chapman, Owen, and Kim Sawchuk. 2015. "Creation-as-Research: Critical Making in Complex Environments." *RACAR: Revue d'art canadienne/ Canadian Art Review* 40 (1): 49–52. http://www.jstor.org/stable/24327426.

Cohen, Daniel J. 2008. "Creating Scholarly Tools and Resources for the Digital Ecosystem: Building Connections in the Zotero Project." *First Monday* 13 (8): n.p. doi:10.5210/fm.v13i8.2233.

Crompton, Constance, Alyssa Arbuckle, Raymond G. Siemens, and the *Devonshire MS* Editorial Group. 2013. "Understanding the Social Edition Through Iterative Implementation: The Case of the Devonshire MS (BL Add MS 17492)." *Scholarly and Research Communication* 4 (3): n.p. http:// src-online.ca/index.php/src/article/viewFile/118/311.

Crompton, Constance, Raymond G. Siemens, and Alyssa Arbuckle, with the INKE Research Group. 2015. "Enlisting 'Vertues Noble & Excelent': Behavior, Credit, and Knowledge Organization in the Social Edition." *Digital Humanities Quarterly* 9 (2): n.p. http://www.digitalhumanities.org/dhq/vol/9/2/000202/000202.html.

Deterding, Sebastian, Dan Dixon, Rilla Khalad, and Lennart E. Nacke. 2011. "From Game Design Elements to Gamefulness: Defining 'Gamification.'" In *Proceedings of the 15th International Academic MindTrek Conference: Envisioning Future Media Environments (MindTrek '11)*, 9–15. New York: ACM. doi:10.1145/2181037.2181040.

Drucker, Johanna. 2003. "Designing Ivanhoe." *TEXT Technology* 12 (2): 19–41. http://texttechnology.mcmaster.ca/pdf/vol12_2_03.pdf.

Eisenstein, Elizabeth L. 1979. *The Printing Press as an Agent of Change: Communications and Cultural Transformations in Early Modern Europe*. Cambridge: Cambridge University Press.

Ellison, Julie. 2008. "The Humanities and the Public Soul." *Antipode* 40 (3): 463–71. doi:10.1111/j.1467-8330.2008.00615.x.

Farland, Maria M. 1996. "Academic Professionalism and the New Public Mindedness." *Higher Education Exchange* Annual: 51–57. http://www.unz.org/Pub/HigherEdExchange-1996q1-00051.

Fernheimer, Janice W., Lisa Litterio, and James Hendler. 2011. "Transdisciplinary ITexts and the Future of Web-scale Collaboration." *Journal of Business and Technical Communication* 25 (3): 322–37. doi:10.1177/1050651911400710.

Fitzpatrick, Kathleen. 2007. "CommentPress: New (Social) Structures for New (Networked) Texts." *Journal of Electronic Publishing* 10 (3): n.p. doi:10.3998/3336451.0010.305.

Galey, Alan, and Stan Ruecker. 2010. "How a Prototype Argues." *Digital Scholarship in the Humanities* (formerly *Literary and Linguistic Computing*) 25 (4): 405–24. doi:10.1093/llc/fqq021.

Gitelman, Lisa. 2006. *Always Already New: Media, History, and the Data of Culture*. Cambridge, MA: MIT Press.

Hendry, David G., J.R. Jenkins, and Joseph F. McCarthy. 2006. "Collaborative Bibliography." *Information Processing & Management* 42 (3): 805–25. doi:10.1016/j.ipm.2005.05.007.

Jay, Gregory. 2012. "The Engaged Humanities: Principles and Practices for Public Scholarship and Teaching." *Journal of Community Engagement and Scholarship* 3 (1): 51–63. http://jces.ua.edu/the-engaged-humanities-principles-and-practices-for-public-scholarship-and-teaching/.

Johns, Adrian. 1998. *The Nature of the Book: Print and Knowledge in the Making.* Chicago: University of Chicago Press.

Kjellberg, Sara. 2010. "I am a Blogging Researcher: Motivations for Blogging in a Scholarly Context." *First Monday* 15 (8): n.p. doi:10.5210/fm.v15i8.2962.

Levy, Michelle. 2010. "Austen's Manuscripts and the Publicity of Print." *ELH* 77 (4): 1015–40. https://muse.jhu.edu/journals/elh/v077/77.4.levy.pdf.

Marshall, Catherine C. 1997. "Annotation: From Paper Books to the Digital Library." In *Proceedings of the Second ACM International Conference on Digital Libraries (Digital Libraries '97)*, 131–40. Philadelphia: ACM. doi:10.1145/263690.263806.

McCarty, Willard. 2005. *Humanities Computing.* New York: Palgrave Macmillan.

McKenzie, D. F. 1999. *Bibliography and the Sociology of Texts.* Cambridge: Cambridge University Press.

Ovsiannikov, Ilia A., Michael A. Arbib, and Thomas H. McNeill. 1999. "Annotation Technology." *International Journal of Human-Computer Studies* 50 (4): 329–62. doi:10.1006/ijhc.1999.0247.

Ratto, Matt, and Robert Ree. 2012. "Materializing Information: 3D Printing and Social Change." *First Monday* 17 (7): n.p. doi:10.5210/fm.v17i7.3968.

Ratto, Matt, Sara Ann Wylie, and Kirk Jalbert. 2014. "Introduction to the Special Forum on Critical Making as Research Program." *The Information Society* 30 (2): 85–95. doi:10.1080/01972243.2014.875767.

Schreibman, Susan, Raymond G. Siemens, and John Unsworth, eds. 2008. *A Companion to Digital Humanities.* Oxford: Blackwell.

Siemens, Raymond G. 2002. "Scholarly Publishing at its Source, and at Present." In *The Credibility of Electronic Publishing: A Report to the Humanities and Social Sciences Federation of Canada*, compiled by Raymond G. Siemens, Michael Best, Elizabeth Grove-White, Alan Burk, James Kerr, Andy Pope, Jean-Claude Guédon, Geoffrey Rockwell, and Lynne Siemens. TEXT Technology 11 (1): 1–128.

Siemens, Raymond G., Meagan Timney, Cara Leitch, Corina Koolen, and Alex Garnett, with the ETCL, INKE, and PKP Research Groups. 2012. "Toward Modeling the Social Edition: An Approach to Understanding the Electronic Scholarly Edition in the Context of New and Emerging Social Media." Digital Scholarship in the Humanities (formerly Literary and Linguistic Computing) 27 (4): 445–61. doi:10.1093/llc/fqs013.

Stillinger, Jack. 1994. Coleridge and Textual Instability: The Multiple Versions of the Major Poems. New York: Oxford University Press.

Unsworth, John. 2000. "Scholarly Primitives: What Methods Do Humanities Researchers Have in Common, and How Might Our Tools Reflect This?" Part of a symposium on Humanities Computing: Formal Methods, Experimental Practice, sponsored by King's College, London, May 13. http://people.virginia.edu/~jmu2m/Kings.5-00/primitives.html.

Wolfe, Joanna. 2002. "Annotation Technologies: A Software and Research Review." Computers and Composition 19 (4): 471–97. doi:10.1016/S8755-4615(02)00144-5.

I. Social Knowledge Creation and Conveyance

Our intention in this first section of the annotated bibliography, *Social Knowledge Creation and Conveyance*, is to provide an environmental scan of the current state of social knowledge creation, conveyance, and production in its many nodes and manifestations. Additionally, this section exposes the relevance of social knowledge creation for current scholarly endeavours and institutions. Many of the books, articles, collections, blog posts, tools, and projects cited inevitably call for institutional transformation and herald a predicted sea change of academic structures in terms of pedagogy, publishing, and production. These calls for reform rely on inherently social structures of creation and engagement. Widespread institutional change is notoriously slow and can be opposed by many, but the shift from models of single authorship and hoarded knowledge to acknowledging networks of shared knowledge creation may indicate a deconstruction of the real or perceived boundary between academic and non-academic communities. The utopian ideal that digital technology can democratize knowledge—and thereby notions of authority and even resources—signals a unique opportunity for social knowledge creation. In this section we aim to synopsize beneficial resources and trends for individuals invested in digital scholarship, academic reform, and cross-community collaboration.

Although certain resources included in this section of the annotated bibliography do derive from science and technology studies or library studies, the entries as a whole reveal a significant bias toward the humanities (and often the digital humanities). Moreover, this section often focuses on scholarly praxis concerns, as evinced by the substantial number of resources relevant to academic publishing or developing digital humanities projects. This bias does not suggest that social knowledge creation practices are limited to humanities scholars, researchers, and practitioners—fascinating and relevant scholarship has been executed in other fields. Rather, the distinct angle speaks to a more specific underlying purpose of this annotated bibliography: to supplement the research of humanities scholars whose interests lie in studying or developing electronic projects and initiatives within the framework of socially produced knowledge. In keeping with the overarching social theme, this annotated bibliography would likely benefit from a comprehensive expansion into other disciplines.

The term "social knowledge creation" can easily become a muddled or catch-all phrase. We consider social knowledge creation as acts of collaboration in order to engage in or produce shared cultural data and/or knowledge products, but in order to more clearly delineate a research scope, the following annotated bibliography has been categorized by specific topoi. The document contains 187 individual entries, which are divided into 12 distinct categories (and collected into an alphabetical list at the end of the section):

1. History of Social Knowledge Production

2. Society, Governance, and Knowledge Construction and Constriction

3. Designing Knowledge Spaces Through Critical Making

4. Social Media Communities, Content, and Collaboration

5. Spatial Humanities and Digital Mapping

6. Crowdsourcing

7. Discipline Formation in the Academic Context

8. Public Humanities

9. The Shifting Future of Scholarly Communication and Digital Scholarship

10. Social Knowledge Creation in Electronic Journals and Monographs

11. Social Knowledge Creation in Electronic Scholarly Editions and e-Books

12. Exemplary Instances of Social Knowledge Construction

13. A Complete Alphabetical List of Selections

The majority of the 187 entries reflect scholarship generated after 2000. The remaining entries include seminal resources such as Michel Foucault's *Discipline and Punish* (1977) and Jerome McGann's *The Textual Condition* (1991). Each category contains from 7 to 42 entries, and entries have been cross-posted between categories when appropriate.[8]

We have arranged the *Social Knowledge Creation and Conveyance* section in a trajectory that moves from the foundational to the abstract to the contemporary,

---

[8] Please note that cross-posted entries are marked with an asterisk (*) after the first instance.

and eventually settles on pertinent instantiations. The first category, "History of Social Knowledge Production," reflects on the narratological basis of contemporary social knowledge creation practices. This category gestures toward three interrelated fields: textual studies (with a focus on the advent of print and its consequences), historical scholarly practices (specifically of scholarly communication, academic journals, and peer review), and media history (concerning the social context of various media and mediums). The conception of knowledge production as plural represents the point of contact between these fields—knowledge is built out of a composite network of players, history, politics, and social contexts. The 14 selected works in this category analyze past practices and instances of social knowledge production in order to more comprehensively understand those of the present.

The second category, "Society, Governance, and Knowledge Construction and Constriction," represents the political and ideological implications of socially creating (or, more often, synthesizing) knowledge. We concede that insofar as it proves rewarding to analyze productive social knowledge construction practices and theories, it is equally pressing to analyze where social knowledge construction is restricted, limited, or ideologically ordered. This category spans a range of subjects from critical theory to sociotechnology studies, and surveys the field of knowledge production from a theoretical standpoint. Additionally, this section examines the politics of group dynamics, and the negotiation and construction of public space—incorporating such foundational publications as Jürgen Habermas's *The Structural Transformation of the Public Sphere* (1991) as well as more contemporary works. Many of the 32 selections directly engage with the digital environment and computational culture. Pertinent questions raised by the selections in this category include: Who constrains knowledge, and how? Through which channels does knowledge flow? And, perhaps most pressing, how does acknowledging the constriction of knowledge influence our present and future decisions regarding policy, law, and society?

"Designing Knowledge Spaces Through Critical Making" surveys scholarship regarding cognizant design, especially in the digital humanities-oriented field of critical making. Critical making integrates the previously disparate fields of more abstract, conceptual critical theory and a sustained commitment to design and building. The 21 selections in this annotated bibliography represent an underlying consensus that since knowledge is frequently created through the collaboration of various individuals, methodologies, and tools, the design of these interactions (or the space where the interactions occur) needs to be examined and implemented critically. As such, many of the selections focus on how to design digital projects and spaces that stimulate social knowledge creation while maintaining certain ethical or discipline-based standards. Articulated through ideas of "learning by doing" and hands-on collaboration, critical making often focuses on social knowledge production with a more literal interpretation of the term "production."

The rise of social media has encouraged a unique proliferation of transnational, national, and local communication and social knowledge creation. "Social Media Communities, Content, and Collaboration," the fourth category, includes scholarship on Web 2.0 practices and the resulting opportunities for social knowledge creation. The polyvocal and democratic undertones of social media present a formidable opportunity for engagement between various groups of people and movements. Although the depth of social media's influence on creating knowledge and culture remains necessarily unclear at this time, many scholars speculate on, encourage, study, and employ social media. The 23 selections in this annotated bibliography range from introducing scholarly social knowledge creation tools to analyzing the inner workings of social knowledge production in popular networks such as Facebook and Wikipedia.

The fifth and sixth categories were newly integrated in the revision of this document from "Social Knowledge Creation: Three Annotated Bibliographies" to its present manifestation. "Spatial Humanities and Digital Mapping" focuses on the practices of present-day spatial humanities research following the shift to a computational mode of spatial inquiry through digital mapping. This transition has resulted in an expanding social element in many branches of the field, and often involves working with large corpora, made possible through the automatization of geoparsing (i.e., the process of linking a place-name with its geographical identifier). From a social angle, users participate actively in the field by collecting live geospatial data on GPS-based devices and uploading them onto a map, such as in the *OpenStreet-Map* community-driven environment. Another major constituent of digital spatial humanities is the availability of large general and field-specific open gazetteers that contain geospatial information and other location details. Many of these gazetteers consist of thousands of entries and are expanding continuously through user contribution. The open availability of geospatial information circumvents its widespread commercialization, and aligns with open access values. Rather than focusing on digital mapping tools, the 12 entries in this category survey theoretical discourse in the field, features of digital mapping, and successful social mapping initiatives. The "Crowdsourcing" category provides examples of the mobilization of social knowledge creation principles in order to assist the development of large-scale research initiatives. Many of the 17 publications in this category consider crowdsourcing through a case study lens, and use specific experiences in crowdsourcing projects to ground their exploration of best practices. Of note, this category features a number of works published by the team behind *Transcribe Bentham*. The included selections masterfully blend critical commentary on three related, yet distinct, areas—crowdsourcing, digital humanities, and public humanities—and are forward-looking in their attention to utilizing past experiences in order to inform and enlighten future researchers.

The seventh category, "Discipline Formation in the Academic Context," focuses on how academic disciplines form socially, with a particular interest in the intersections between discipline formation and social knowledge creation. Ideally, academic practices and institutions evolve perpetually in order to better serve students, communities, and scholarly practitioners alike. We can often assess the history and current state of the academy through its scholarly communication and discipline formation habits. In keeping with an underlying historical bent, the 29 selected texts span the last three decades of academic writing. The more contemporary resources often tend toward graduate training in humanities programs. The entries range from particular studies of specific areas, such as first year English composition requirements in Canada and the development of ballooning as a field, to wider-lens views of contemporary scholarly institutions at large. Certain selections draw from other disciplines and are intended to reflect on similarities and differences between disciplines. Overall, the entries aim to provide a sense of the varied practices involved in contemporary discipline formation, with an eye to humanities methods.

"Public Humanities" is another new category in this iteration of the annotated bibliography. This category focuses on problem-based scholarship in which members of the university and community engage in discourse that addresses relevant problems in the community, and work on practical ways of solving them. Public humanities now occupy a more central position in institutional practices as a response to the persistent request for universities to be more engaged with community life and enhancement. The public sphere has critized higher education institutions for indulging in isolated, highly specialized, discipline-specific areas at the cost of civic engagement, while still relying on public funding. Many scholars find the humanities especially suited to engage the suggested behavioural change due to the field's disposition to critically discern large, complex problems. The seven selections in this category address the shift toward a more publicly engaged scholarship and the role that the humanities could occupy in the public sphere. It provides models to overcome current limitations in university policy, while integrating an appropriate infrastructure for growth.

The ninth, tenth, and eleventh categories are centred explicitly on academic concerns of social knowledge creation in the digital realm. Category 9, "The Shifting Future of Scholarly Communication and Digital Scholarship," raises a series of questions: What is the role of the humanities in social knowledge production? How can academics harness new tools and modes of scholarship to engage productively with each other as well as with other members of the public? How can the humanities actively reflect on and proactively repurpose the history of scholarly communication? How can digital practices, including publishing, foster social knowledge creation from within the academy? What is the role of open access in social knowledge creation? The 42 selections attend to these questions and branch out further, from rethinking literary criticism to imagining future digital libraries to politicizing the digital humanities. The most stimulating and notable intersections occur when the social and the scholarly overlap. In categories 10 and 11, we consider how current scholarly communication preferences of scholars and editorial teams have led to the thoughtful development of digital scholarly publications. The 14 selections of category 10, "Social Knowledge Creation in Electronic Journals and Monographs," consider how journals and monographs can enable and enact social knowledge practices in the online sphere. In many instances, authors meditate on how these actions can benefit scholarship and scholars both inside and outside the academy. In other cases, authors advocate for further integration of the democratic, user-based interactions and productions encouraged by the rise and popularity of Web 2.0 practices. In still other entries, authors ruminate on the history of the academic journal and apply this knowledge to the current state of scholarly communication. Taken as whole, the selections introduce the nuanced conversation surrounding contemporary journal and monograph production. Category 11, "Social Knowledge Creation in Electronic Scholarly Editions and e-Books," acknowledges how the form and function of digital scholarly editions and e-books have evolved in parallel with the Internet itself. Simultaneously, digital scholarly editions and e-books carry forward and reflect bibliographic theories, often concerning the inherent sociality of texts. The 24 selections consider many far-reaching issues, including how editors can harness the allowances of the digital realm to best represent the social text, and how authors can facilitate social knowledge creation via electronic publication. Many selections also reflect on content creators integrating already-existent social knowledge production practices within their projects. Perhaps most noteworthy, certain selections ask what is lacking in digital editions, and how they might be improved.

The twelfth category, "Exemplary Instances of Social Knowledge Construction," provides a sampler of particularly relevant social knowledge creation tools and platforms. In our conception of the term, a social knowledge creation tool is a usable technology that encourages the collaborative work of multiple individuals in a networked, digital environment. Furthermore, a social knowledge creation tool supports the active generation of information or knowledge in an ethos of sharing, contact, and openness. The 11 selections in this category reflect a range of practices and social knowledge creation tools, from community bibliography to folksonomy tagging to collaborative annotation. Finally, a complete alphabetical list of selections concludes this section of the annotated bibliography.

Social Knowledge Creation and Conveyance moves consciously among what may at first appear to be disparate schools of thought. With purposely broad strokes, the document comments on the productive or beneficial qualities of social knowledge production, and should be considered as a supplemental resource for those interested in studying, initiating, or participating in social knowledge creation. Readers can expect to gain a nuanced sense of the history, stakes, opportunities, and conversation surrounding contemporary social knowledge practices, especially in the digital realm.

1. History of Social Knowledge Production

Bazerman, Charles. 1991. "How Natural Philosophers Can Cooperate: The Literary Technology of Coordinated Investigation in Joseph Priestley's *History and Present State of Electricity* (1767)." In *Textual Dynamics of the Professions: Historical and Contemporary Studies of Writing in Professional Communities,* edited by Charles Bazerman and James Paradis, 13–44. Madison: University of Wisconsin Press.
Bazerman studies the role of early literature reviews through a thorough discussion and analysis of Joseph Priestley's *The History and Present State of Electricity* (1767). If, as Bazerman argues, literature reviews represent potent sites of knowledge-sharing and dissemination in a community, then Priestley's volume represents the first literature review, since it details the history of electricity research and experiments. Priestley created a comprehensive, open-ended document that summarized the accepted state of the field as well as anomalies, discrepancies, and failures. Bazerman applauds Priestley for his active service in democratizing and disseminating knowledge.

Biagioli, Mario. 2002. "From Book Censorship to Academic Peer Review."
   Emergences **12 (1): 11–45. doi:10.1080/1045722022000003435.**
Biagioli details the historical and epistemological shifts that have led to the
academic peer review system as it is now known. Contrary to its contemporary role, peer review began as an early modern disciplinary technique
closely related to book censorship and required for social and scholarly
certification of institutions and individuals alike. The rise of academic journals shifted this constrained and royally-mandated position; no longer a
self-sustaining system of judgment and reputation dictated by a small group
of identified and accredited professionals, (often blind) peer review now focuses on disseminating knowledge and scholarship to the wider community.
Biagioli also states that journals have moved from officially representing
specific academic institutions to being community owned and operated, as
responsibilities, duties, and readership are now dispersed among a group of
like-minded scholars.

Burke, Peter. 2000. *A Social History of Knowledge: From Gutenberg to*
   Diderot. **Cambridge: Polity Press.**
Burke expands on the various agents and elements of social knowledge production, with a specific focus on intellectuals and Europe in the early modern period (until ca. 1750). He argues that knowledge is always plural, and
that various knowledges concurrently develop, surface, intersect, and play.
Burke relies on sociology, including the work of Emile Durkheim, and critical
theory, including the work of Michel Foucault, as a basis on which to develop
his own notions of social knowledge production. He acknowledges that the
church, scholarly institutions, government, and the printing press have all
had a significant effect on knowledge production and dissemination—often
affirmatively, but occasionally through restriction or containment. Furthermore, Burke explores how both "heretics" (humanist revolutionaries) and
more traditional academic structures developed the university as a knowledge institution.

_____ **2012.** *A Social History of Knowledge II: From the Encyclopédie to*
   Wikipedia. **Cambridge: Polity Press.**
Burke develops his research from the first volume (*A Social History of Knowledge: From Gutenberg to Diderot*) by expanding his scope from the early modern
period into the twentieth century. He continues to rely on certain foundational notions for this volume: knowledge is plural and varied; knowledge
is produced by various institutions and conditions instead of solely by individuals; and the social production of knowledge is intrinsically connected to
the economic and political environments in which it develops. As with the

first volume, Burke focuses mainly on academic knowledge, with brief forays into other forms or sites of knowledge.

Eagleton, Terry. 2010. "The Rise of English." In *The Norton Anthology of Theory and Criticism*, edited by Vincent B. Leitch, 2140–46. New York: W.W. Norton.
Eagleton argues that the development of English literature was an ideological strategy used, beginning in the mid-nineteenth century, as a form of suppression and control to educate lower classes only "enough" to keep them subservient. English literature, moreover, was actually scorned and primarily directed at women when first introduced as a field of university study. Eagleton concludes that literature "is an ideology" (2140) due to its historical role in social development and nation-building in England and elsewhere.

Fjällbrant, Nancy. 1997. "Scholarly Communication—Historical Development and New Possibilities." In *Proceedings of the IATUL Conferences*. West Lafayette, IN: Purdue University Libraries e-Pubs. http://docs.lib.purdue.edu/cgi/viewcontent.cgi?article=1389&context=iatul.
In order to study the widespread transition of scholarly communication from print to electronic formats, Fjällbrant details the history of the scientific journal. Academic journals had emerged in seventeenth-century Europe, and the first of these, the *Journal des Sçavans*, was published in 1665 in Paris. The first learned societies formed at this time—the Royal Society in London and the Académie des Sciences in Paris—were primarily concerned with the dissemination of knowledge, and the scholarly journal developed out of a desire by researchers to share their findings with others in a cooperative forum. Following the lead of the Royal Society, some of whose members had read the *Journal des Sçavans*, other societies established similar serial publications. Although there were other contemporaneous forms of scholarly communication, including the letter, the scientific book, the newspaper, and the cryptic anagram system, the journal emerged as a primary source of scholarly communication. It met the needs of various stakeholders: the general public, booksellers and publishers, libraries, authors who wished to make their work public and claim ownership, the scientific community invested in reading and applying other scientists' findings, and academic institutions that required metrics for evaluating faculty.

Gitelman, Lisa. 2006. *Always Already New: Media, History, and the Data of Culture*. Cambridge, MA: MIT Press.
Gitelman relates media history, with a focus on contextual social processes, in order to examine human experience, communication, and cultural history.

She argues that media are plural, socially recognized communication structures that evolve with surrounding publics. Gitelman rejects contemporary notions of media as a singular, ubiquitous force—"The Media." Instead, she examines two contrasting technologies, the phonograph and the Internet, envisioning media as active, multiple, historical subjects. Gitelman briefly extends her argument into the materiality of media subjects, digital versus non-digital textual materiality, and the necessary omnipresence of both form and content.

Jagodzinski, Cecile M. 2008. "The University Press in North America: A Brief History." *Journal of Scholarly Publishing* **40 (1): 1–20. doi:10.1353/scp.0.0022.**
Jagodzinski describes the history of the North American university press, beginning with the first presses at Cornell and Johns Hopkins Universities, which debuted in the nineteenth century. From the beginning, the primary function of the university press was considered to be the dissemination of knowledge. Twentieth-century growth in the number of colleges and universities led to a corresponding growth in the number of university presses, and the Association of American University Presses (AAUP) was formally established in the mid-1930s. As is well known, the last quarter of the twentieth century heralded major systemic changes and obstacles, and the university press was not immune to these challenges. Jagodzinski discusses in detail how university presses have responded, pragmatically and creatively, to the (largely financial) issues burdening contemporary scholarly communication.

Johns, Adrian. 1998. *The Nature of the Book: Print and Knowledge in the Making.* **Chicago: University of Chicago Press.**
Johns, a self-professed historian of printing, seeks to reveal a social history of print: a new, more accurate exploration of how print, and thereby knowledge, developed. His account of print includes acknowledging the labours of those actually involved with printing, as well as their contemporary understandings and anxieties surrounding print and publication. With a distinct focus on the history of science, Johns explores the social apparatus and construction of print, as well as how print has been used socially. Notably, Johns constructs his argument in firm opposition to Elizabeth Eisenstein's earlier work on print culture (1979). He argues that there is no singular "print culture," as such; rather, there are various print cultures that are all local in character. For Johns, the wide-ranging influence of print is manifold, multiple, and not implicit in a deterministic cause and effect relationship with any single historical factor or trigger.

Liu, Alan. 2013. "From Reading to Social Computing." In *Literary Studies in the Digital Age: An Evolving Anthology*, edited by Kenneth M. Price and Raymond G. Siemens, n.p. New York: MLA Commons. https://dlsanthology.commons.mla.org/from-reading-to-social-computing/.

Liu presents an impressive short history of both social computing and literary theory. He develops the argument that literary scholars must take social computing seriously, as it is the current mode of cultural and personal expression. Liu suggests that literary scholars engage with social computing through two distinct methodologies: those of the social sciences and of the digital humanities. As he argues, social computing must be considered not only as an *object* of literary study, but as a *practice* of literary study.

Manovich, Lev. 2001. *The Language of New Media*. Cambridge, MA: MIT Press.

Manovich distills both abstract and assumed theories concerning the history and present state of computing and media. In doing so, he attempts to contextualize, categorize, and develop a relevant vocabulary of new media. Concurrently, Manovich explains the technical development of new media, situating it in the twentieth-century media trajectory—with one eye to cinema and the other to print. His contextualization reveals how new media and previous media mutually define and inform each other. Manovich discusses the transformations that cause the digital computer to act as a cultural processor and a "universal media machine" (4). He further defines new media by enumerating five principles: automation, numerical code, access, variability, and transcoding. Of note, Manovich proposes the opposition of database and narrative due to differences in form and linearity.

Siemens, Raymond G. 2002. "Scholarly Publishing at its Source, and at Present." In *The Credibility of Electronic Publishing: A Report to the Humanities and Social Sciences Federation of Canada*, compiled by Raymond G. Siemens, Michael Best, Elizabeth Grove-White, Alan Burk, James Kerr, Andy Pope, Jean-Claude Guédon, Geoffrey Rockwell, and Lynne Siemens. *TEXT Technology*, 11 (1): 1–128.

Siemens's introduction to this report focuses on the rethinking of scholarly communication practices in light of new digital forms. He meditates on this topic through the framework of *ad fontes*—the act, or conception, of going to the source. As he argues, scholars should look at the source or genesis of scholarly communication. For Siemens, the source goes beyond the seventeenth-century inception of the academic print journal to include less formal ways of communicating and disseminating knowledge—i.e., verbal exchanges, epistolary correspondence, and manuscript circulation. In this way, scholars can look past the popular, standard academic journal and into a future of scholarly communication that productively involves varied scholarly traditions and social knowledge practices.

Streeter, Thomas. 2010. "Introduction." In *The Net Effect: Romanticism, Capitalism, and the Internet*, 1–16. New York and London: New York University Press.
Through a distinctly sociological method, Streeter analyzes the connections between computing, the rise of the Internet, capitalism, and social life. Instead of framing his examination through the Internet's effect on society, Streeter looks at how the Internet has been socially constructed, and at its role in myriad complex historical, personal, and political networks. Rather than speculating on its possible future, he questions why and how the Internet was built. Moreover, Streeter discredits essentialist conceptions of technology and the Internet; he articulates that various historical and cultural contexts have fostered the openness of the Internet's networked state.

Turner, Fred. 2006. *From Counterculture to Cyberculture: Stewart Brand, the Whole Earth Network, and the Rise of Digital Utopianism.* Chicago: University of Chicago Press.
Turner's sociohistorical narrative of the development of the Internet makes the argument that the counterculture movements of the 1960s—specifically those under the leadership of Stewart Brand and the Whole Earth Network—have played an integral role in both the principles and practices of contemporary personal computing. He posits that the New Communalists (those who flocked to communes in the late 1960s and early 1970s) assisted and influenced the widespread network of computing as we know it today through their embrace of cybernetics and a technology-based ideology. Turner elaborates on the social construction of modern computing, as well as on how computing influenced numerous American social groups, movements, and citizens in both abstract and tangible ways.

2. Society, Governance, and Knowledge Construction and Constriction

Althusser, Louis. 1971. "Ideology and Ideological State Apparatuses (Notes Towards an Investigation)." In *Lenin and Philosophy and Other Essays*, translated by Ben Brewster, 127–86. New York: Monthly Review Press.
Althusser describes the form and function of ideology, and how it dictates experience and knowledge via Ideological State Apparatuses (ISAs). ISAs

include the church (the "religious ISA"), family, school, union, law, culture, political system, and communication infrastructure. Repressive State Apparatuses (RSAs), on the other hand, include more overtly violent institutions such as the police and the army. ISAs constitute subjects, and thus experience, through ritual and practice. As they are omnipresent institutions, ISAs dictate knowledge production: subjects both constitute and are constituted by ISAs. Althusser contends that the school is the prime contemporary instantiation of the ISA; the school maintains an ideological infrastructure through the training of children into ideological subjectivity, thereby reproducing the conditions of production.

Ang, Ien. 2004. **"Who Needs Cultural Research?" In** *Cultural Studies and Practical Politics: Theory, Coalition Building, and Social Activism,* **edited by Pepi Leystina, 477–83. New York: Blackwell.**

Ang ruminates on the current relationship between cultural studies, the university, the public, and society at large. She argues that not only do individuals benefit from cultural studies work, but they in fact rely on this sort of work to navigate, comprehend, and meaningfully contribute to an increasingly complex world. Ang advocates for the detachment of cultural studies from corporate-based funding, as she worries that these sorts of partnerships will, by catering to popular will and interest, falsely skew and inadequately represent the field of cultural research. Ang asserts that social knowledge production must be supported by a knowledge infrastructure that holistically approaches the study and creation of culture.

Bailey, Moya Z. 2011. **"All the Digital Humanists Are White, All the Nerds Are Men, But Some of Us Are Brave."** *Journal of Digital Humanities* **1 (1): n.p. http://journalofdigitalhumanities.org/1-1/all-the-digital-humanists-are-white-all-the-nerds-are-men-but-some-of-us-are-brave-by-moya-z-bailey/.**

Bailey situates herself in a critical conversation on the racialized and gendered terminology of nerddom, and by extension, as she argues, of the digital humanities. Bailey asserts that individuals who identify as being on the margins of traditional academia will often find themselves at the borders of digital humanities as well. She argues that if we can open the field and engage those often left on margins (women, disabled individuals, people of colour), an entirely new set of theoretical questions and directions will become viable. As case studies, Bailey offers projects and activism initiatives that carry out this objective.

Balsamo, Anne. 2011. Introduction: "Taking Culture Seriously in the Age of Innovation." In *Designing Culture: The Technological Imagination at Work*, 2–25. Durham, NC: Duke University Press.

Balsamo studies the intersections of culture and innovation, and acknowledges the unity between the two modes ("technoculture"). She argues that technological innovation should seriously recognize culture as both its inherent context and a space of evolving, emergent possibility, as innovation necessarily alters culture and social knowledge creation practices. Balsamo introduces the concept of the "technological imagination"—the innovative, actualizing mindset. She also details a comprehensive list of truisms about technological innovation, ranging from considering innovation as performative, historically constituted, and multidisciplinary to acknowledging design as a major player in cultural reproduction, social negotiation, and meaning-making. Currently, innovation is firmly bound up with economic incentives, and the profit-driven mentality often obscures the social and cultural consequences and implications of technological advancement. As such, Balsamo calls for more conscientious design, education, and development of technology, and a broader vision of the widespread influence and agency of innovation.

*Biagioli, Mario. 2002. "From Book Censorship to Academic Peer Review." *Emergences* 12 (1): 11–45. doi:10.1080/1045722022000003435.

Biagioli details the historical and epistemological shifts that have led to the academic peer review system as it is now known. Contrary to its contemporary role, peer review began as an early modern disciplinary technique closely related to book censorship and required for social and scholarly certification of institutions and individuals alike. The rise of academic journals shifted this constrained and royally-mandated position; no longer a self-sustaining system of judgment and reputation dictated by a small group of identified and accredited professionals, (often blind) peer review now focuses on disseminating knowledge and scholarship to the wider community. Biagioli also states that journals have moved from officially representing specific academic institutions to being community owned and operated, as responsibilities, duties, and readership are now dispersed among a group of like-minded scholars.

Benkler, Yochai. 2003. "Freedom in the Commons: Towards a Political Economy of Information." *Duke Law Journal* 52 (6): 1245–76. http://scholarship.law.duke.edu/dlj/vol52/iss6/3.

Benkler analyzes the pervasive social influence of the Internet, with a focus on the economic and political changes affected by the rise and ubiquity of digital spaces, networks, and action. He argues that the Internet has caused two new social phenomena to occur: "nonmarket production" (production by an individual without intention to generate profit) and "decentralized production" (production that occurs outside of the sanctioned centres of industry). In turn, these phenomena facilitate new opportunities to pursue democracy, individual freedom, and social justice. The forms of production incited by the Internet permit individuals and communities to gain control over their work, means of production, and networks of relations, and consequently to garner more influence. Benkler concludes by rallying readers to take advantage of the opportunities the digital environment boasts in order to build more just and democratic social, economic, and political systems.

Berry, David M. 2012. "The Social Epistemologies of Software." *Social Epistemology: A Journal of Knowledge, Culture and Policy* 26 (3-4): 379–98. doi:10.1080/02691728.2012.727191.

Berry analyzes how code and software increasingly develop, influence, and depend on social epistemology or social knowledge creation. He discusses the highly mediated "computational ecologies" (379) that individuals and non-human actors inhabit, and argues that we need to become more aware of the role these computational ecologies play in daily social knowledge production. Berry analyzes two case studies to support his argument: the existence of web bugs or user activity trackers, and the development of lifestreams, real-time streams, and the quantified self. For Berry, the increasing embrace of and compliance with potentially insidious data collecting via the Internet and social media needs to be addressed.

Bijker, Wiebe E., and John Law. 1992. "General Introduction." In *Shaping Technology/Building Society: Studies in Sociotechnical Change*, edited by Wiebe E. Bijker and John Law, 1–14. Cambridge, MA: MIT Press.

In the introduction to this collection, Bijker and Law develop the overarching theme of the included essays: the social construction, context, and relations of technology, especially concerning design and inception. They argue that technologies are never isolated or prefabricated, but are generated out of a set of varying circumstances and actors. Bijker and Law acknowledge various relevant theories from sociotechnology to constructivism to the social history of technology. Notably, the authors focus on the idea that "it might have been otherwise" (4), and employ the phrase as a guiding mantra for both their inquiry and the collection at large.

Bourdieu, Pierre. 1993. "The Field of Cultural Production, or: The Economic World Reversed." In *The Field of Cultural Production: Essays on Art and Literature*, edited and translated by Randal Johnson, 29–73. New York: Columbia University Press.

Bourdieu dictates his vision of the field of cultural production as inherently socially mediated, from production to reception. He concedes that since all cultural artifacts exist as symbolic objects—"manifestation[s] of the field as a whole" (38)—one cannot study a cultural artifact without acknowledging the material and symbolic production of the work. Furthermore, the field of cultural production, although in some ways autonomous, is contained within both the field of power and the field of class relations. In fact, in what seems to be reverse logic, the more autonomous a field of cultural production becomes, the less power the field has in regard to the fields of power and class relations; autonomy, for Bourdieu, represents an increased reliability on an internal system of logic and success, and therefore a further distancing from other fields.

Burdick, Anne, Johanna Drucker, Peter Lunenfeld, Todd Presner, and Jeffrey Schnapp. 2012. "The Social Life of the Digital Humanities." In *Digital_Humanities*, 73–98. Cambridge, MA: MIT Press.

Burdick et al. focus on the social aspects and impacts of digital humanities. The authors argue that the digital humanities, by nature, encompass academic and social spaces that discuss issues beyond technology alone. Key issues include open access, open source publications, the emergence of participatory Web and social media technologies, collaborative authorship, crowdsourcing, knowledge creation, influence, authorization, and dissemination. Burdick et al. also consider the role of digital humanities in public spaces, beyond the siloed academy. The authors address these expansive issues through an oscillating approach of explanation and questioning. While the diversity of the topics in this chapter is substantial, the authors knit the arguments together under the broad theme of social engagement.

Chun, Wendy Hui Kyong. 2004. "On Software, or the Persistence of Visual Knowledge." *Grey Room* 18: 26–51. doi:10.1162/1526381043320741.

Chun re-evaluates the supposed transparency of software, and instead focuses on the blackboxing, abstraction, and causal pleasure that define contemporary computing and programming. She reinscribes software as akin to ideology: intangible but present, persuasive, subject/user- producing, and capable of rendering the visible invisible and vice versa. Concurrently, Chun studies the gendered history of computation and programming, observing how contemporary accounts of this history mask some major female players and early entrepreneurs. Furthermore, Chun argues, the mechanization of computers shifted power relations and ostensibly wrote women out of the computing and programming narrative. Chun concludes that we must acknowledge, interrogate, and criticize the obscuring tendencies of software in order to avoid submitting to its ideological nuances.

*Eagleton, Terry. 2010. "The Rise of English." In *The Norton Anthology of Theory and Criticism,* edited by Vincent B. Leitch, 2140–46. New York: W.W. Norton.

Eagleton argues that the development of English literature was an ideological strategy used, beginning in the mid-nineteenth century, as a form of suppression and control to educate lower classes only "enough" to keep them subservient. English literature, moreover, was actually scorned and primarily directed at women when first introduced as a field of university study. Eagleton concludes that literature "is an ideology" (2140) due to its historical role in social development and nation-building in England and elsewhere.

Edwards, Charlie. 2012. "The Digital Humanities and Its Users." In *Debates in the Digital Humanities,* edited by Matthew K. Gold, 213–32. Minnesota: University of Minnesota Press. http://dhdebates. gc.cuny.edu/debates/text/31.

Edwards examines digital humanities as a system, and traces the history of the "user" (i.e., scholar) in this discipline. He argues that many digital humanities tools see limited uptake because "traditional" humanists are less enthusiastic about computational methods, or have not heard of or do not understand what the digital humanities are. The article is structured as a response to "Stuff Digital Humanists Like: Defining Digital Humanities by its Values" (2010), in which Tom Scheinfieldt argues that digital humanists do not just make use of tools and values from the Internet, but also work with and contribute to the Internet. Edwards suggests that this process has yet to be realized, and that digital humanities lack a central set of agreed-upon values. Edwards compares digital humanities Twitter activity to a stock ticker, and writing in the digital humanities to a network, noting that navigating, communicating, and contributing to the discipline can be challenging. He calls for the digital humanities to offer an understandable, loose, accessible framework of the discipline for outsiders.

Flanders, Julia. 2012. "Time, Labor, and 'Alternate Careers' in Digital Humanities Knowledge Work." In *Debates in the Digital Humanities,* **edited by Matthew K. Gold, 292–308. Minneapolis: University of Minnesota Press. http://dhdebates.gc.cuny.edu/debates/text/26.**
Flanders investigates alternative careers, especially those that exist in relation to digital humanities projects, and compares them to traditional humanities, especially in terms of time management and salary. Through a set of examples from her own career, Flanders points to how working in traditional humanities, from graduate school and all the way up to tenure track, is measured in qualitative terms; this makes the time and reward for this work limitless and merges the boundary between academic and personal life. She compares this to working on digital projects in alternative careers, including project management skills and quantifiable time set per hour/project to be spent on a task, resulting in a more concrete schedule. Flanders concludes by an appeal to traditional humanities to improve their system of evaluation of time spent on projects and to reward students, staff, and faculty members for publishing their work.

Fraser, Nancy. 1990. "Rethinking the Public Sphere: A Contribution to the Critique of Actually Existing Democracy." *Social Text* **(25/26): 56–80. http://www.jstor.org/stable/466240.**
Fraser expands upon and critiques Jürgen Habermas's concept of the public sphere. She argues that Habermas's idea of the public sphere is a useful conceptual resource for overcoming problems of political participation in modern societies. Fraser points out a foundational internal irony in Habermas's theory: a discourse touting accessibility, rationality, and suspension of status is itself a form of hierarchy. As she demonstrates, the relationship between the public and the private is much more complicated than Habermas intimates, and, in general, revisionist historiography reveals a much darker bourgeois public sphere than the one that emerges from Habermas's work. Central to Habermas's theory is the right to open accessibility. Fraser draws attention to the non-realization of this ideal in history—specifically when it comes to participation from marginal groups. Fraser's argument demonstrates how Habermas's theory is inadequate for critiquing the interactions of late capitalist societies.

Freeman, Jo. 1972. "The Tyranny of Structurelessness." *The Second Wave* **2 (1): n.p. http://www.jofreeman.com/joreen/tyranny.htm.**
While structurelessness first emerged as a counter-reaction to an over-structured society, Freeman asserts that structurelessness has now become a system of organization in its own right. Freeman's central argument is that

there is no such thing as a structureless group, and that any group of people that come together, of any nature and for any length of time, will inevitably structure itself in some way. Freeman argues that structureless groups, while possible in theory, are impossible in practice. Many elitist organizations, Freeman argues, hide behind the idea of structurelessness. Therefore, because of the impossibility of structurelessness in practice, Freeman argues that structure must be made formal and explicit in order to avoid the formation of an elitist or exclusive group. The conditions that promote productive group dynamics do not occur naturally in many large groups and, consequently, it is challenging for big groups to get things done. Freeman rounds out her article with a list of practical principles that are key to democratic structuring.

Foucault, Michel. 1977. *Discipline and Punish: The Birth of the Prison.* Translated by Alan Sheridan. London: Allen Lane and Penguin Books.

Foucault details the complex history of contemporary discipline and punishment structures and networks. He maintains that various forces of normalization, along with a pervasive carceral system, are responsible for knowledge formation, the social body, and modern notions of punishment, justice, legality, and delinquency. Of note, in the penultimate section, "Discipline," Foucault identifies specific elements utilized in order to maintain docile subjects through disciplinary methods: place (via enclosure, partitioning, and delineating space based on rank); time (via timetables, notions of efficiency, and the temporal mechanization of the body); mechanic efficiency (via command, chronological series, and reducing the body to a part of a larger machine); normalization (via differentiation, hierarchy, homogenization, and exclusion); examination (via objectification, documentation, and making an individual a "case"); and surveillance (via spatial partitioning, panoptic structures, and the intertwining of surveillance and economy). Foucault concludes that no individual is outside of the system; the carceral network wherein everyone resides creates so-called "delinquents."

Graff, Gerald. 1987. *Professing Literature: An Institutional History.* Chicago: University of Chicago Press.

Graff thoroughly details the history of twentieth-century English literature studies in America. He argues that many of the issues in contemporary academia can be traced to an overall method of patterned isolationism in a department. Due to intellectual or discipline-based conflicts, various isolated fields of thought and practitioners prevail. Conflicts are neither acknowledged nor attended to, but rather overlooked by a general attitude of inclusion and comprehensiveness. As a result, divergent schools of thought never engage in conversation or debate, and all practitioners are endowed with silos in which they can effectively ignore their intellectual opponents. The self-perpetuating lack of interconnectedness and collaboration in English departments has negatively affected their overall scholarship and success. Furthermore, Graff contends that the conflict between schools of thought (classicism, New Criticism, critical theory, and now, perhaps, digital humanities) should be taught to students in order to contextualize and lend meaning to their literary education. Graff presents the above arguments as an introduction to a comprehensive historical explanation of how literary studies evolved as a discipline.

Habermas, Jürgen. 1991. "Introduction: Preliminary Demarcation of a Type of Bourgeois Public Sphere." In *The Structural Transformation of the Public Sphere*, translated by Thomas Burger with the assistance of Frederick Lawrence, 1–26. Cambridge, MA: MIT Press.

In this foundational article, Habermas examines the history and various definitions of "public." Beginning with ancient Greece and Rome, Habermas points to the polis and bios politikos as common (although not necessarily physical) spaces for free citizens to engage in public life. For him, the conception of a public sphere emerged during the Renaissance and came to fruition in the eighteenth century. Habermas traces the shifting relationship between public and private across this evolutionary history. Our modern understanding of the division between public and private, according to Habermas, appeared with the rise of the capitalist economy and the popularization of the press—a new forum in which the public could share and generate criticism. While Habermas suggests that the public sphere presents an inclusive opportunity for public engagement, he limits the population of eligible citizens to educated males.

Haraway, Donna. 1990. "A Cyborg Manifesto: Science, Technology, and Socialist Feminism in the Late Twentieth Century." In *Simians, Cyborgs, and Women: The Reinvention of Nature*, 149–81. New York: Routledge.

Haraway advocates for the new social relations of science and technology through criticizing essentialist feminism, Marxism, and anti-science and technology politics simultaneously. She argues that by embodying the form of the nebulous, ungendered, unboundaried cyborg figure, science and technology can be harnessed for productive political means. Haraway contends that ideological opposition to technology only reinforces the futility of movements that follow notions of hierarchies and origin stories. The fluid, hybrid cyborg represents an opportunity for the marginalized to constitute knowledge production by participating in new forms of social relations afforded by technology.

Heidegger, Martin. 1982. "The Question Concerning Technology." In *The Question Concerning Technology and Other Questions*, translated with an introduction by William Lovitt, 3–35. New York: Harper Perennial.
Heidegger contends that we must consider both the "essence" of technology and our role as humans concerning technology: we do not control technology, nor are we technology, nor does technology control us. Rather, technology is better understood as a revealer, as a mediator, or as that which performs "en-framing." En-framing denotes a calling into being (or else a contextualization) by technology. Recognizing technology's true essence as an enframer—instead of as a tool, an oppressive other, or as fate—increases our awareness of existence.

Introna, Lucas D., and Helen Nissenbaum. 2000. "Shaping the Web: Why the Politics of Search Engines Matters." *The Information Society* 16 (3): 169–85.
According to Introna and Nissenbaum, search engines are frequently biased in their findings, and thus in their representation of what is available on the Internet. The authors argue that this tendency bears serious implications, as the digital realm is often perceived and promoted as a democratic, empowering space. Introna and Nissenbaum detail the various processes that promote "findability" on the Internet. Furthermore, they caution against the commercialization of search engines, lest they become authoritative arbiters of the digital divide. Introna and Nissenbaum conclude by reminding their readers that public digital acts are more than simply technical matters—they often bear political implications as well, especially concerning issues of access and capital.

Lessig, Lawrence. 2004. *Free Culture: How Big Media Uses Technology and the Law to Lock Down Culture and Control Creativity*. New York: Penguin.
Lessig argues that the interests of a select (corporate) few have increasingly regulated contemporary American society by legislating the Internet with intellectual property and piracy laws. According to Lessig, this regulation defeats traditional American ideals of democracy and free culture, and constrains social knowledge creation and important cultural and intellectual advances. Lessig respects the concept of copyright and intellectual property,

as such—he takes issue with the hyperregulation and restriction of the Internet and, consequently, individuals. Moreover, Lessig demonstrates how all culture industries have "stolen" from previous individuals, art forms, and media. Paradoxically, the same industries persecute individuals for practising intellectual or creative theft.

Liu, Alan. 2004. *The Laws of Cool: Knowledge Work and the Culture of Information*. Chicago: University of Chicago Press.

Liu interweaves two distinct threads in *The Laws of Cool*. He traces the history and ethos of "cool" (culture, trends, popularity, etc.) as well as postindustrial cool: the flux of cool knowledge work. Liu examines how the humanities can contribute to and survive in the new postindustrial cool, corporate landscape. Liu's sources and interests are widespread; he cites modernist design theory, Lev Manovich's database narrative, and everything from the Gayaki tribe to William Gibson's *Agrippa*. He concludes that the humanities are necessary to keep the corporation humane and informed of the history of its own practices; the humanities, in turn, must learn to negotiate the current cool cultural climate in order to remain relevant and effective.

McPherson, Tara. 2012. "Why are the Digital Humanities So White? or Thinking the Histories of Race and Computation." In *Debates in the Digital Humanities*, edited by Matthew K. Gold, 139–60. Minnesota: University of Minnesota Press. http://dhdebates.gc.cuny.edu/debates/text/29.

McPherson explores how to knit together distinct issues in the humanities and the digital humanities. She argues that the ethnically homogenous computational culture that came of out World War II caused the current, fraught intersection of race and technology. McPherson narrates two fragments from history in the 1960s to illustrate her argument: computer scientists working to develop UNIX/MULTICS, and the assassination of Malcolm X. She argues that while these two events are parallel in time and are deeply related, they appear siloed because they attract separate audiences. McPherson urges that race and even post-structuralism be put in conversation with technology as fundamental factors in the shaping of the discipline. She does not argue that technological innovations consciously encode racism, but rather that information responds to racial justice in many registers. McPherson encourages scholars to educate themselves on the machines and networks that shape our lives, and to acknowledge the role of computers as coders of culture.

Nowviskie, Bethany. 2012. "Evaluating Collaborative Digital Scholarship (or, Where Credit is Due)." *Journal of Digital Humanities* 1 (4): n.p. http://journalofdigitalhumanities.org/1-4/evaluating-collaborative-digital-scholarship-by-bethany-nowviskie/.

Nowviskie begins this article (developed out of a conference talk) by identifying a key disjuncture in the discipline of digital humanities: while collaboration is touted as a hallmark of digital humanities scholarship, it is glossed over in conversations about tenure and promotion. Nowviskie argues that the tenure and promotion process is ill-fitted to assessing digital humanities research because it relies on the fiction that only final outputs are scholarship. Digital scholarship, Nowviskie asserts, is rarely "done," and that complicates our traditional notions of assessment. She argues that acknowledging project collaborators fairly can contribute to imaginative production, enthusiastic promotion, and committed preservation—three vital characteristics of collaborative scholarship. As we open scholarship up to new kinds of work, we must also accept new kinds of peer review and definitions of authorship. Nowviskie concludes with six basic principles of evaluation to reconfigure traditional humanities principles.

Ross, Anthony, and Nadia Caidi. 2005. "Action and Reaction: Libraries in the Post 9/11 Environment." *Library and Information Science Research* 27 (1): 97–114. doi:10.1016/j.lisr.2004.09.006.

Ross and Caidi study the significantly shifting roles and responsibilities of North American libraries following 9/11 and the subsequent enactment of legislation (the USA PATRIOT Act). Traditionally public information institutions, libraries have become increasingly regulated regarding confidentiality, patron privacy, and intellectual freedom, as well as access to and handling of government information. Ross and Caidi also explore reactions to the substantial change in legislation. These reactions reveal libraries' willingness and ability to effect political change when faced with intrusive restrictions on their traditional roles in sharing and promoting knowledge.

Siemens, Lynne. 2009. "It's a Team if You Use 'Reply All': An Exploration of Research Teams in Digital Humanities Environments." *Digital Scholarship in the Humanities* (formerly *Literary and Linguistic Computing*) 24 (2): 225–33. doi:10.1093/llc/fqp009.

Siemens identifies a singular contrast between traditional humanities research and digital humanities research: while traditionally the humanities as a discipline functioned as predominantly solo research efforts, the digital humanities involves various individuals with a wide spectrum of skills working together. Siemens argues that the collaborative nature of academic research communities, especially in the humanities, has been understudied. She fills that gap by examining the results of interviews conducted on the topics of teams, team-based work experiences, and team research preparation. The interviewees identify both benefits and challenges of team research, including rich interactions, relationship building with potential for future projects, communication challenges, funding, and team member retention. To conclude, Siemens articulates a list of five essential practices: (i) deliberate action by each team member; (ii) deliberate action by the project leader; (iii) deliberate action by the team; (iv) deliberate training; and (v) balance between digital and in-person communication.

*Streeter, Thomas. 2010. "Introduction." In *The Net Effect: Romanticism, Capitalism, and the Internet*, 1–16. New York and London: New York University Press.

Through a distinctly sociological method, Streeter analyzes the connections between computing, the rise of the Internet, capitalism, and social life. Instead of framing his examination through the Internet's effect on society, Streeter looks at how the Internet has been socially constructed, and at its role in myriad complex historical, personal, and political networks. Rather than speculating on its possible future, he questions why and how the Internet was built. Moreover, Streeter discredits essentialist conceptions of technology and the Internet; he articulates that various historical and cultural contexts have fostered the openness of the Internet's networked state.

*Turner, Fred. 2006. *From Counterculture to Cyberculture: Stewart Brand, the Whole Earth Network, and the Rise of Digital Utopianism*. Chicago: University of Chicago Press.

Turner's sociohistorical narrative of the development of the Internet makes the argument that the counterculture movements of the 1960s—specifically those under the leadership of Stewart Brand and the Whole Earth Network—have played an integral role in both the principles and practices of contemporary personal computing. He posits that the New Communalists (those who flocked to communes in the late 1960s and early 1970s) assisted and influenced the widespread network of computing as we know it today through their embrace of cybernetics and a technology-based ideology. Turner elaborates on the social construction of modern computing, as well as on how computing influenced numerous American social groups, movements, and citizens in both abstract and tangible ways.

Vaidhyanathan, Siva. 2002. "The Content-Provider Paradox: Universities in the Information Ecosystem." *Academe* 88 (5): 34–37. doi:10.2307/40252219.
Vaidhyanathan warns against the increasing corporatization of American universities and other knowledge institutions. He argues that universities have begun to commodify knowledge, and that this tactic will eventually lead to the dissolution of the university as a credible source of education. Unfortunately, Vaidhyanathan does not offer an alternative model through which universities can address widespread funding and budget cuts. Nevertheless, taking a similar approach to that of Willard McCarty in *Humanities Computing*, Vaidhyanathan reminds his readers that education is not simply information, and should not be treated (or sold) as such.

Williams, George H. 2012. "Disability, Universal Design, and the Digital Humanities." In *Debates in the Digital Humanities*, edited by Matthew K. Gold, 202–12. Minneapolis: University of Minnesota Press. http://dhdebates.gc.cuny.edu/debates/text/44.
Williams acknowledges a major limitation of digital humanities: the field has not addressed the needs of people who are differently abled, especially within the context of preservation and accessibility of digital information. For Williams, this is crucial to resolve, as many digital humanities projects are federally funded and their materials must be made openly accessible to the public by law. Williams points out that contemporary web standards and practices can accommodate the needs (and devices) of many. He proposes implementing universal design in the creation of digital resources, and provides examples of different projects that work to make the digital humanities a more inclusive field. Williams concludes by acknowledging that this beneficial direction would ensure that digital resources are useful and usable by a wide array of people now and in the future.

3. Designing Knowledge Spaces Through Critical Making

Arbuckle, Alyssa, and Alex Christie, with the ETCL, INKE, and MVP Research Groups. 2015. "Intersections Between Social Knowledge Creation and Critical Making." *Scholarly and Research Communication* 6 (3): n.p. http://src-online.ca/index.php/src/article/view/200.
Arbuckle and Christie outline the practices of digital scholarly communication (moving research production and dissemination online), critical making (producing theoretical insights by transforming digitized heritage materials), and social knowledge creation (collaborating in online environments to produce shared knowledge products). In addition to exploring these

practices and their principles, the authors argue that combining these activities engenders knowledge production chains that connect multiple institutions and communities. Highlighting the relevance of critical making theory for scholarly communication practice, Arbuckle and Christie provide examples of theoretical research that offer tangible products for expanding and enriching scholarly production.

*Balsamo, Anne. 2011. Introduction: "Taking Culture Seriously in the Age of Innovation." In *Designing Culture: The Technological Imagination at Work*, 2–25. Durham, NC: Duke University Press.
Balsamo studies the intersections of culture and innovation, and acknowledges the unity between the two modes ("technoculture"). She argues that technological innovation should seriously recognize culture as both its inherent context and a space of evolving, emergent possibility, as innovation necessarily alters culture and social knowledge creation practices. Balsamo introduces the concept of the "technological imagination"—the innovative, actualizing mindset. She also details a comprehensive list of truisms about technological innovation, ranging from considering innovation as performative, historically constituted, and multidisciplinary to acknowledging design as a major player in cultural reproduction, social negotiation, and meaning-making. Currently, innovation is firmly bound up with economic incentives, and the profit-driven mentality often obscures the social and cultural consequences and implications of technological advancement. As such, Balsamo calls for more conscientious design, education, and development of technology, and a broader vision of the widespread influence and agency of innovation.

*Bijker, Wiebe E., and John Law. 1992. "General Introduction." In *Shaping Technology/Building Society: Studies in Sociotechnical Change*, edited by Wiebe E. Bijker and John Law, 1–14. Cambridge, MA: MIT Press.
In the introduction to this collection, Bijker and Law develop the overarching theme of the included essays: the social construction, context, and relations of technology, especially concerning design and inception. They argue that technologies are never isolated or prefabricated, but are generated out of a set of varying circumstances and actors. Bijker and Law acknowledge various relevant theories from sociotechnology to constructivism to the social history of technology. Notably, the authors focus on the idea that "it might have been otherwise" (4), and employ the phrase as a guiding mantra for both their inquiry and the collection at large.

Chapman, Owen, and Kim Sawchuk. 2015. "Creation-as-Research: Critical Making in Complex Environments." *RACAR: Revue d'art canadienne / Canadian Art Review* **40 (1): 49–52. http://www.jstor. org/stable/24327426.**
Chapman and Sawchuk discuss the importance of "creation-as-research" and argue that it is not highly valued among other practices. Creation-as-research draws on research outcomes focused on process and material aspects. The authors argue that this modality of research creation requires further reflection, as it is increasingly incorporated into university courses. By comparing projects that share similar features, they consider the essential concepts and productive ironies and tensions of creation-as-research. Chapman and Sawchuk conclude that creation-as-research and critical making are crucial for innovative collectives.

Drucker, Johanna. 2009. "From Digital Humanities to Speculative Computing." In *SpecLab: Digital Aesthetics and Projects in Speculative Computing*, **3–18. Chicago: University of Chicago Press.**
Drucker locates speculative computing as a more critical extension and reflection of digital humanities practices. Knowledge is interpretive and fluid, and thereby conflicts with many computational principles (discrete objects, interoperability, objectivity) that form the basis for the application side of digital humanities. Drucker thus situates herself, and speculative computing at large, as the interrogator of digital humanities standards and normalized practices—based on concepts of knowledge as complex experience versus knowledge as mere information. Notably, Drucker calls for an increased awareness of design as a purposeful mediator instead of as an objective deliverer of information. She concludes by ruminating on models as dynamic, interpretive interventions invaluable for speculative computing at large.

_____ 2011. "Humanities Approaches to Interface Theory." *Culture Machine* **12: 1–20. http://www.culturemachine.net/index.php/cm/ article/viewArticle/434.**
In Drucker's humanities theory of interface, she argues that the interface is the predominant site of cognition in digital spaces and requires cognizant, intellectual design. Drucker's theory is predicated on interface design that considers the constitution of a subject, not the expected activities of a user; on graphical reading practices and frame theory; on constructivist approaches to cognition, and on integrating multiple modes of humanities interpretation. She argues for a humanities approach to interface theory that integrates different forms of reading and analysis in order to allow readers to recognize the relations of the dynamic space between environments

and cognitive events. Furthermore, while avoiding a descent into screen essentialism, Drucker insists that studying electronic reading practices must be focalized through studying graphical user interfaces, as GUIs constitute reading (and thus the reading subject, or "subject of interface" [3]).

_____ 2012. "Humanistic Theory and Digital Scholarship." In *Debates in the Digital Humanities*, edited by Matthew K. Gold, 85–95. Minneapolis: University of Minnesota Press. http://dhdebates. gc.cuny.edu/debates/text/34.

Drucker argues that humanities intervention is pertinent at the level of design for digital projects and incentives. Without humanistic theories, Drucker contends, knowledge, events, experience, and data are at the risk of being flattened and reified. Frequently, humanities theory is not integrated into digital scholarship and development because computer science techniques and theories (mechanization, automation, independent/isolated items) remain at odds with those of the humanities (fluidity, interpretation, and interconnectedness). These barriers must be overcome in order to comprehensively and reflexively create and share knowledge.

Fisher, Caitlin. 2015. "Mentoring Research-Creation: Secrets, Strategies, and Beautiful Failures." *RACAR: Revue d'art canadienne / Canadian Art Review* 40 (1): 46–49. http://www.jstor.org/stable/24327425.

Fisher discusses the importance and potential of research-creation in the graduate education context. She argues that in order to reach a more innovative level of academic production, faculties need to be open to risks and to accept that failure is part of the process. Fisher discusses doctoral projects and expresses her admiration for making and creation. She explains that faculty members often treat this kind of student project as impossible to understand, in contrast to their own work. She concludes that there are audiences who are highly interested in such an approach, hence the importance of appreciating failure and encouraging students to take risks.

Hart, Jennefer, Charlene Ridley, Faisal Taher, Corina Sas, and Alan J. Dix. 2008. "Exploring the Facebook Experience: A New Approach to Usability." In *Proceedings of the 5th Nordic Conference on Human-Computer Interaction (NordiCHI08)*, 471–74. New York: ACM.

In the framework of user experience design, Hart, Ridley, Sas, Taher, and Dix examine a selection of users' reactions to the popular social networking website Facebook. The authors put forth the idea that previous standards of evaluating digital environments need to be reimagined for our current technological moment to privilege user experience. Their findings indicate an overall positive reaction to Facebook despite the site's meeting only two out of the ten traditional usability guidelines. The authors call for a more holistic approach to design that pays heed to the pleasurable social knowledge creation experience of many individuals as they participate on social networking sites such as Facebook.

Jessop, Martyn. 2008. "Digital Visualization as a Scholarly Activity." *Digital Scholarship in the Humanities* **(formerly** *Literary and Linguistic Computing***) 23 (3): 281–93. doi:10.1093/llc/fqn016.**
Jessop argues that digital visualization deserves to be taken seriously as scholarly work by fellow academics. Digital visualization creates an opportunity for new knowledge production, as well as increased visual literacy and diverse intellectual practices. Jessop comments at length on the form and function of digital visualization, and its role in relation to the humanities at large. He reflects that from an academic standpoint, digital visualization is frequently criticized as not scholarly, or not scholarly enough. To overcome these limiting assumptions, Jessop advocates for the adherence to a set of standards (in this article, he promotes the London Charter) in order to validate digital visualization and to ensure that a lasting debate shapes and maintains the practice and its concurrent knowledge creation.

Latour, Bruno. 2009. "A Cautious Prometheus? A Few Steps Towards a Philosophy of Design (with Special Attention to Peter Sloterdijk)." In *Networks of Design: Proceedings of the 2008 Annual International Conference of the Design History Society*, **edited by Fiona Hackne, Jonathan Glynne, and Viv Minto, 2–10. Boca Raton, FL: Universal Publishers.**
Latour meditates on the form and function of the term "design," and proposes a more comprehensive vision for the practice. He suggests that design practitioners focus more fully on drawing together, modelling, or simulating complexity—more inclusive visions that incorporate contradiction and controversy. Latour argues that we are living in an age of design (or redesign) instead of a revolutionary modernist era of breaking with the past and making everything new. Increasingly, design encapsulates various other acts, from arrangement to definition, from projecting to coding. Consequently, the possibilities and instances for design grow exponentially. For Latour, the concept of an age of design predicates an advantageous condition defined by humility and modesty (because it is not foundational or construction-based); a necessary attentiveness to details and skillfulness; a focus on purposeful development (or on the meaning of what is being designed); thoughtful

remediation; and an ethical dimension (exemplified through the good design versus bad design binary).

Liu, Alan. 2012. "Where is Cultural Criticism in the Digital Humanities?" In *Debates in the Digital Humanities*, edited by Matthew K. Gold, 490–510. Minneapolis: University of Minnesota Press. http://dhdebates. gc.cuny.edu/debates/text/20.

Liu surveys the state of the digital humanities in relation to the humanities at large. He argues that, thus far, digital humanities projects have often lacked the self-reflexivity and cultural criticism necessary for the ethical development of humanistic projects—thereby denying the digital humanities a real or full position in the humanities. Because the digital humanities avoid cultural criticism, they frequently become subservient or merely instrumental to the humanities as a whole, functioning as either a moneymaker or tech support. Liu claims that the digital humanities could deconstruct the hierarchy by becoming both self-reflexive and invaluable, thereby leading the humanities into the academic future.

McCarty, Willard. 2005. *Humanities Computing.* New York: Palgrave Macmillan.

McCarty examines the field of humanities computing and explores both its limitations and potential. He frames much of his exploration through the mantra that digital humanities can be much more than merely "convenient vending machines for knowledge" (6); the focus must be shifted from automation and delivery to the possibilities for new knowledge creation through digital humanities practices. To this end, McCarty celebrates the tendency toward modelling and manipulation. Drawing heavily on Clifford Geertz's model of/model for theory (and privileging the "model for" concept), McCarty explores how models and unfinished prototypes can be productive spaces of work, knowledge, and play. Models provide invaluable information when they dysfunction, either through inexplicable successes or failures. Of note, he incorporates Martin Heidegger's concept of manipulating the world through technology.

McGillivray, David, Gayle McPherson, Jennifer Jones, and Alison McCandlish. 2016. "Young People, Digital Media Making and Critical Digital Citizenship." *Leisure Studies* 35 (6): 724–38. doi:10.1080/0261 4367.2015.1062041.

McGillivray, McPherson, Jones, and McCandlish explore the debates on digital media in the twenty-first century. They emphasize the role of media in the educational setting by supplying critical insights on the Digital

Commonwealth project, a non-profit organization that supports the creation, management, and dissemination of relevant materials in Massachusetts cultural heritage institutions. They suggest that the prosumer (producer and consumer of online content) challenges the formality of schools as a space for learning. However, they point out that schools have been resisting the leisure opportunities in digital environments by preventing students from using social tools for educational purposes. The authors base their study of the Digital Commonwealth project on recent academic, policy, and practice-focused international work, including that of David Gauntlett (*Making is Connecting: The Social Meaning of Creativity, from DIY and Knitting to YouTube and Web 2.0*, 2011) and Larry Johnson et al. (*Horizon Report Europe: 2014 Schools Edition*, 2014) in order to display how digital media allows individuals to navigate between different worlds and environments. The authors conclude that the social web allows simultaneous leisure and learning, and that educators and students must invest in it in order to produce beneficial outcomes rather than just consume content passively.

Ramsay, Stephen, and Geoffrey Rockwell. 2012. "Developing Things: Notes Toward an Epistemology of Building in the Digital Humanities." In *Debates in the Digital Humanities*, edited by Matthew K. Gold, 75– 84. Minneapolis: University of Minnesota Press.
Ramsay and Rockwell take up the "your database/prototype is an argument" conversation (notably championed by Lev Manovich and Willard McCarty). They assert that taking building as seriously as scholarly work could productively dismantle or realign the focus of the humanities from its predominantly textual bent. Ramsay and Rockwell advocate for installing the user, reader, or subject at the level of building. Through this socially minded conceptual and physical shift, some of the abstractions and black boxing that render digital humanities tools theoretically insufficient could be avoided or amended.

Ratto, Matt. 2011a. "Critical Making: Conceptual and Material Studies in Technology and Social Life." *The Information Society* 27 (4): 252– 60. doi:10.1080/01972243.2011.583819.
Ratto briefly but effectively describes his engagement with critical making as a scholarly practice. For Ratto, critical making integrates conceptual critical theory and practical, hands-on material work, with the aim of furthering comprehension of the role of technology in social life. Ratto reflects on his own experiences and varying degrees of success in practising critical making with different groups of scholars. Of note, Ratto concludes that personal

investment influences the connection between lived experience (making) and developing critical perspectives on social issues.

_____ 2011b. "Open Design and Critical Making." In *Open Design Now: Why Design Cannot Remain Exclusive*, edited by Bas van Abel, Lucas Evers, Roel Klaassen, and Peter Troxler, n.p. Amsterdam: BIS Publishers. http://opendesignnow.org/index.php/article/critical-making-matt-ratto/.

Ratto situates his conception and practice of critical making within the context of open design. He argues that critical making encapsulates one of the major tenets of open design: reconnecting morality and materiality. Ratto addresses sociotechnological issues through a constructivist engagement with scholarly research and pedagogy. He contends that open design is necessary—both practically and theoretically—for the continued success of the critical making movement. Critical making relies substantially on the ethos, as well as the support, of the open design community. Open design, in turn, should embrace critical making as a scholarly pursuit aimed at studying (as well as criticizing) accepted social practices.

Ratto, Matt, and Robert Ree. 2012. "Materializing Information: 3D Printing and Social Change." *First Monday* 17 (7): n.p. doi:10.5210/fm.v17i7.3968.

Ratto and Ree argue that 3D printing is emerging as a socially transformative technology with a positive effect on creative activity. They focus on the process and practice of 3D printing, and discuss its technical aspects as well as its scholarly and entrepreneurship influences. The authors propose that 3D printing will result in various changes, and that the role of government and the creative sector has to extend beyond current efforts and initiatives in this regard. Ratto and Ree conclude by calling for additional research in the field.

Ratto, Matt, Sara Ann Wylie, and Kirk Jalbert. 2014. "Introduction to the Special Forum on Critical Making as Research Program." *The Information Society* 30 (2): 85–95. doi:10.1080/01972243.2014.875767.

Ratto, Wylie, and Jalbert claim that critical making should be considered a research field, since its practices are diverse, far-reaching, and inherently valuable. They believe that representational and material approaches can provide a framework for researchers involved in similar work. Critical making can offer further contextualization and expand relevance of academic work. The authors refer to Ratto (2011a, annotated above) and Ratto and Hockema ("FLWR PWR: Tending the Walled Garden" [2009]) to suggest that critical making as a research program explores and connects conceptual critique and material practice. After discussing the distinctive practices of critical making, they conclude by suggesting that the issue at hand is to place making within the broader conceptual structure of knowledge work.

Vetch, Paul. 2010. "From Edition to Experience: Feeling the Way towards User-Focused Interfaces." In *Electronic Publishing: Politics and Pragmatics*, edited by Gabriel Egan, 171–84. New Technologies in Medieval and Renaissance Studies 2. Tempe, AZ: Iter Inc., in collaboration with the Arizona Center for Medieval and Renaissance Studies.

Vetch explores the nuances of a user-focused approach to scholarly digital projects, arguing that the prevalence of Web 2.0 practices and standards requires scholars to rethink the design of scholarly digital editions. For Vetch, editorial teams need to shift their focus to questions concerning the user. For instance, how will users customize their experience of the digital edition? What new forms of knowledge can develop from these interactions? Moreover, how can rethinking the interface design of scholarly digital editions promote more user engagement and interest? Vetch concludes that a user-focused approach is necessary for the success of scholarly publication in a constantly shifting digital world.

Wylie, Sara Ann, Kirk Jalbert, Shannon Dosemagen, and Matt Ratto. 2014. "Institutions for Civic Technoscience: How Critical Making is Transforming Environmental Research." *The Information Society* 30 (2): 116–26. doi:10.1080/01972243.2014.875767.

Wylie, Jalbert, Dosemagen, and Ratto conduct two case studies to explore the relationship between academic and public formations of scientific science, which they call "civic technoscience." They argue that civic technoscience gives the public access and opportunity, in order to question expert knowledge production, as well as allowing them to create credible, public science. By referring to scholars in the field, Wylie et al. suggest that communities in environmental health and social justice are trying to adopt knowledge making as a formal research interface. They conclude by challenging the academy to support civic technoscience. The authors encourage academic researchers to provide access to their labs, classes, and tools, since the public in fact funds most university equipment.

4. Social Media Communities, Content, and Collaboration

*Berry, David M. 2012. "The Social Epistemologies of Software." *Social Epistemology: A Journal of Knowledge, Culture and Policy* **26 (3–4): 379–98. doi:10.1080/02691728.2012.727191.**
Berry analyzes how code and software increasingly develop, influence, and depend on social epistemology or social knowledge creation. He discusses the highly mediated "computational ecologies" (379) that individuals and non-human actors inhabit, and argues that we need to become more aware of the role these computational ecologies play in daily social knowledge production. Berry analyzes two case studies to support his argument: the existence of web bugs or user activity trackers, and the development of lifestreams, real-time streams, and the quantified self. For Berry, the increasing embrace of and compliance with potentially insidious data collecting via the Internet and social media needs to be addressed.

Bolter, Jay David. 2007. "Digital Media and Art: Always Already Complicit?" *Criticism* **49 (1): 107–18. doi:10.1353/crt.2008.0013.**
Bolter examines the contemporary theoretical conversation surrounding new media, and identifies a blind spot with regard to social media and computing. He argues that although many contemporary scholars and artists study, discuss, or create digital media, none of them takes into account the cultural significance of social media and computing. Bolter explicitly focuses his examination on the work of Lisa Gitelman, Marie-Laure Ryan, and Johanna Drucker, but also engages with other theorists, including N. Katherine Hayles and Lev Manovich. For Bolter, the transgressive identity and group formation that characterize social media and computing enact the historical goal of the avant garde: to disrupt the boundaries between art, creation, and everyday life.

Boot, Peter. 2012. "Literary Evaluation in Online Communities of Writers and Readers." *Scholarly and Research Communication* **3 (2): n.p. http://src-online.ca/index.php/src/article/view/77/90.**
Boot analyzes the Dutch website Verhalensite ("story site," www.verhalensite.com) in order to understand the mechanics of rating and reputation in online writing communities. He demonstrates that a literary work is rated in online environments against a large number of other works, noting that the literary evaluation itself takes place in public (online). In addition, user communication takes place on the same site, where one has access to the literary works and their evaluations. To conduct his analysis, Boot refers to quid pro quo (reciprocal commenting), comment words, and network analysis. He

concludes that the process of online literary evaluation varies from that of print literature: formalized literary institutions are mostly absent from online communities, whereas social media sites give more power to the reader in determining the reputation of a work. Boot adds that online writing communities share similarities with other online communities, and calls for a comparison of the mechanisms of reputation.

*Burdick, Anne, Johanna Drucker, Peter Lunenfeld, Todd Presner, and Jeffrey Schnapp. 2012. "The Social Life of the Digital Humanities." In *Digital_Humanities*, 73–98. Cambridge, MA: MIT Press.
Burdick et al. focus on the social aspects and impacts of digital humanities. The authors argue that the digital humanities, by nature, encompass academic and social spaces that discuss issues beyond technology alone. Key issues include open access, open source publications, the emergence of participatory Web and social media technologies, collaborative authorship, crowdsourcing, knowledge creation, influence, authorization, and dissemination. Burdick et al. also consider the role of digital humanities in public spaces, beyond the siloed academy. The authors address these expansive issues through an oscillating approach of explanation and questioning. While the diversity of the topics in this chapter is substantial, the authors knit the arguments together under the broad theme of social engagement.

Cao, Qilin, Yong Lu, Dayong Dong, Zongming Tang, and Yongqiang Li. 2013. "The Roles of Bridging and Bonding in Social Media Communities." *Journal of the American Society for Information Science and Technology* 64 (8): 1671–81. doi:10.1002/asi.22866.
Cao, Lu, Dong, Tang, and Li develop a theoretical model investigating the contribution of bonding (social networks among homogeneous groups) and bridging (social networks among heterogeneous groups) to the individual and collective well-being of virtual communities, through information exchange. They argue that bridging and bonding have positive implications on information quality but not quantity, also noting that information quality is more critical than information quantity after a disaster. They situate their work within the social capital theory, referring to Nan Lin ("Social Networks and Status Attainment," 1999), Pierre Bourdieu and Loïc J.D. Wacquant (*An Invitation to Reflexive Sociology*, 1992), and Mamata Bhandar, Shan-Ling Pan, and Bernard C.Y. Tan ("Towards Understanding the Roles of Social Capital in Knowledge Integration: A Case Study of a Collaborative Information Systems Project," 2007). The authors conclude that bonding has an impact on bridging, and that both have a positive impact on information quality.

Clement, Tanya. 2011. "Knowledge Representation and Digital Scholarly Editions in Theory and Practice." *Journal of the Text Encoding Initiative* **1: n.p. doi:10.4000/jtei.203.**

Clement reflects on scholarly digital editions as sites of textual performance, wherein the editor lays down and privileges various narrative threads for the reader to pick up and interpret. She underscores this theoretical discussion with examples from her own work with the digital edition *In Transition: Selected Poems by the Baroness Elsa von Freytag-Loringhoven*, as well as TEI and XML encoding and the Versioning Machine. Clement details how editorial decisions shape the social experience of an edition. By applying John Bryant's theory of the fluid text to her own editorial practice, she focuses on concepts of various textual performances and meaning-making events. Notably, Clement also explores the idea of the social text network. She concludes that the concept of the network is not new to digital editions; nevertheless, conceiving of a digital edition as a network of various players, temporal spaces, and instantiations promotes fruitful scholarly exploration.

Cohen, Daniel J. 2008. "Creating Scholarly Tools and Resources for the Digital Ecosystem: Building Connections in the Zotero Project." *First Monday* **13 (8): n.p. doi:10.5210/fm.v13i8.2233.**

Cohen details how the Zotero project exemplifies both a Web 2.0 and a traditional scholarly ethos. He conceptualizes Zotero as a node in an interconnected digital ecosystem that builds bridges instead of hoarding information. Zotero is a widely used, open source, community-based bibliography tool. It exists on top of the browser as an extension, has maintained an API since its inception, and boasts comprehensive user features. As an easy-to-use collaborative tool, Zotero acts as both an effective scholarly resource and a facilitator of social knowledge creation.

Flanders, Julia. 2009. "The Productive Unease of 21st-Century Digital Scholarship." *Digital Humanities Quarterly* **3 (3): n.p. http://www. digitalhumanities.org/dhq/vol/3/3/000055/000055.html.**

Flanders discusses the role of the digital humanities in relation to the more conventional humanities, and characterizes the digital humanities as possessing a sort of "productive unease": anxiety concerning medium, institutional structures of scholarly communication, and representation. This anxiety is productive insofar as it brings into clearer focus previously unremarked upon biases in the traditional humanities. Moreover, digital tools and practices present more and different challenges. Of note, Flanders recognizes social software and media as tackling some of these anxiety-provoking issues, and acknowledges digital humanities projects that also strive in the same direction.

*Hart, Jennefer, Charlene Ridley, Faisal Taher, Corina Sas, and Alan J. Dix. 2008. "Exploring the Facebook Experience: A New Approach to Usability." In *Proceedings of the 5th Nordic Conference on Human-Computer Interaction (NordiCHI08)*, 471–74. New York: ACM.

In the framework of user experience design, Hart, Ridley, Sas, Taher, and Dix examine a selection of users' reactions to the popular social networking website Facebook. The authors put forth the idea that previous standards of evaluating digital environments need to be reimagined for our current technological moment to privilege user experience. Their findings indicate an overall positive reaction to Facebook despite the site's meeting only two out of the ten traditional usability guidelines. The authors call for a more holistic approach to design that pays heed to the pleasurable social knowledge creation experience of many individuals as they participate on social networking sites such as Facebook.

Hendry, David G., J. R. Jenkins, and Joseph F. McCarthy. 2006. "Collaborative Bibliography." *Information Processing & Management* 42 (3): 805–25. doi:10.1016/j.ipm.2005.05.007.

Hendry, Jenkins, and McCarthy provide an overview of the type of bibliographies published on the web today, and extend the traditional view to encompass participatory practices. By providing a conceptual model for the infrastructure of these practices, the authors demonstrate the process of producing and supporting these collections, both on a theoretical level and through a case study. The ideal result of these participatory policies would involve an environment with collaborative decision-making, a visible workflow and collective shaping of it, and audience discussions. However, they conclude that the realization of this model would require a great investment in systems development, and is not yet sustainable.

Inversini, Alessandro, Rogan Sage, Nigel Williams, and Dimitrios Buhalis. 2015. "The Social Impact of Events in Social Media Conversation." In *Information and Communication Technologies in Tourism 2015*, edited by Iis Tussyadiah and Alessandro Inversini, 283–94. Lugano, Switzerland: Springer International Publishing.

Inversini, Sage, Williams, and Buhalis discuss how the growth of social media allows for analysis of real time discussions that happen on online platforms. The authors study the impact of real time discussions on the discourse of a given event. Their research objectives are to understand the extent to which social media has a propensity to facilitate socially motivated discussion, and to establish a correlation between the platform users' centrality in the network and their effect on such discussions. To do that, the authors investigate socially motivated discussions taking place on Twitter about the Glastonbury Music Festival (UK). Inversini et al. conclude that it is important to study a community of interests on social media platforms, since socially motivated discussion generates more impact on the event in question. They note the necessity of examining online narratives of events in order to understand engagement levels with social causes.

Kittur, Aniket, and Robert E. Kraut. 2008. "Harnessing the Wisdom of the Crowds in Wikipedia: Quality Through Coordination." In *Proceedings of the 2008 ACM Conference on Computer Supported Cooperative Work (CSCW 08)*, 37–46. New York: ACM.

Kittur and Kraut study the correlation between the number of editors on a Wikipedia page and the quality of that page's content. Significantly, they argue that an increased number of editors on a given page will prove productive only if some sort of coordination apparatus is in place. Articles are even more successful, content-wise, if a small group of experts manages the majority of the work. This argument runs counter to the crowdsourcing ethos of Wikipedia, which dictates that, generally, the more editors at work, the better the quality of the article. The authors argue, however, that a smaller group of editors working under a semi-authoritative organizational system facilitates peer-to-peer communication—a benefit that is often lost when large groups of uncoordinated individuals are involved.

Kirschenbaum, Matthew. 2012. "Digital Humanities As/Is a Tactical Term." In *Debates in the Digital Humanities*, edited by Matthew K. Gold, 415–28. Minneapolis: University of Minnesota Press. http:// dhdebates.gc.cuny.edu/debates/text/48.

For Kirschenbaum, digital humanities should be considered as a tactical term because of its notable role as a means instead of simply as an end. He argues that social media environments and interactions highlight this tactical nature. For instance, social networks and blogs (particularly Twitter) offer a space for digital humanists to engage in alternative professional interaction and dialogue. Kirschenbaum indicates, however, that Twitter's significance exceeds the sheer presence of digital humanist users; the digital humanities community is in fact established through social media's tendency to build reputations and status, metrically indicate influence, and aggregate information and like-minded individuals. Thus, while accepted scholarly channels and institutions continue to represent the digital humanities in a more traditional sense, the community's tactical, online existence promotes constant change and alternative forms of professional clout.

Kjellberg, Sara. 2010. "I am a Blogging Researcher: Motivations for Blogging in a Scholarly Context." *First Monday* **15 (8): n.p. doi:10.5210/fm.v15i8.2962.**
Kjellberg explores the incentives behind blogging in an academic context. To do so, she focuses on twelve researchers' activities, as well as the specific functionalities of the blogging environment. According to the researchers, the social function of communication and knowledge dissemination inherent in blogging results in a feeling of connectedness. Kjellberg identifies some of the main blogging functionalities as expressing opinions, keeping up-to-date, and disseminating research. Blogging does not necessarily provide researchers with direct career advantages, but persists nonetheless due to an interest in developing the practice, as well as in the possibility of reaching multiple audiences.

Liu, Alan. 2011. "Friending the Past: The Sense of History and Social Computing." *New Literary History: A Journal of Theory and Interpretation* **42 (1): 1–30. doi:10.1353/nlh.2011.0004.**
Liu identifies media-induced sociality in oral, written, and digital culture. He proceeds to analyze Web 2.0 and social computing practices, and concludes that Web 2.0 lacks a sense of history, despite its intricately interconnected state. Liu attributes this state to two concurrent historical shifts: a social move from one-to-many to many-to-many, and a temporal shift from straightforward conceptions of time into the contemporary conception of instantaneous and simultaneous temporality. Reflexively, Liu argues that conceiving of time in this new instantaneous/simultaneous framework may ideologically proprietize the Internet and allow for ownership of social practices by organizations such as Facebook, Twitter, and Google. As such, Liu opts for a more traditional sense of temporality and history characterized by narratological linear time. He cites the social network system of his Research-oriented Social Environment (RoSE) project as a platform that integrates history with Web 2.0 infrastructure and allowances.

*_____ 2013. "From Reading to Social Computing." In *Literary Studies in the Digital Age: An Evolving Anthology*, **edited by Kenneth M. Price and Raymond G. Siemens, n.p. New York: MLA Commons. https://dlsanthology.commons.mla.org/from-reading-to-social-computing/.**
Liu presents an impressive short history of both social computing and literary theory. He develops the argument that literary scholars must take social computing seriously, as it is the current mode of cultural and personal expression. Liu suggests that literary scholars engage with social computing

through two distinct methodologies: those of the social sciences and of the digital humanities. As he argues, social computing must be considered not only as an *object* of literary study, but as a *practice* of literary study.

Manovich, Lev. 2012. "Trending: The Promises and the Challenges of Big Social Data." In *Debates in the Digital Humanities*, edited by Matthew K. Gold, 460–75. Minneapolis: University of Minnesota Press. http:// dhdebates.gc.cuny.edu/debates/text/15.

Manovich elaborates on the possibilities and limitations of performing humanities research with Big Data. He asserts that although Big Data can be incredibly instructive and useful for humanities work, certain significant roadblocks impede this project. These roadblocks include the fact that only social media companies have access to relevant Big Data; user-generated content is not necessarily authentic, objective, or representative; certain analysis of Big Data requires a level of computer science expertise that humanities researchers do not typically possess; and Big Data is not synonymous with "deep data," the type of data procured through intense, long-term study of subjects. Nevertheless, Manovich looks forward to a future where humanists can overcome these boundaries and integrate Big Data with their research aspirations and projects.

Hart, William, and Terry Marsh. 2014. "Social Media Research Foundation." In *Encyclopedia of Social Media and Politics*, edited by Kerric Harvey, 3:1173–74. Thousand Oaks, CA: Sage.

Hart and Marsh introduce the Social Media Research Foundation, with a focus on its structure and objectives: understanding the many aspects of social media. They suggest that the foundation's interdisciplinary research and development of social media tools and data sets holds significant implications for fields such as national politics and international relations. The authors claim that NodeXL (one of the foundation's open tools for analyzing social media), and the data one can generate using it, can offer researchers an in-depth view of social media users and their followers.

Mrva-Montoya, Agata. 2012. "Social Media: New Editing Tools or Weapons of Mass Distraction?" *Journal of Electronic Publishing* 15 (1): 1–24. doi:10.3998/3336451.0015.103.

Mrva-Montoya discusses the effect and usages of social media in the editorial profession. She argues that, when used appropriately, social media have a positive impact on editors, enabling them to sustain professional relationships, garner information and responses quickly and easily, and build their reputation and status. In contrast, social media have a negative impact on editors when usage becomes overly time consuming, distracting, revealing, or overbearing. By harnessing the productive effects of social media, editorial professionals can proactively manage their careers and success.

Pfister, Damien Smith. 2011. "Networked Expertise in the Era of Many-to-Many Communication: On Wikipedia and Invention." *Social Epistemology: A Journal of Knowledge, Culture and Policy* 25 (3): 217–31. doi:10.1080/02691728.2011.578306.
Pfister argues that Wikipedia is a prime example and facilitator of contemporary many-to-many communication structures and the resultant changing nature of knowledge production. He advocates for many-to-many communication because it disrupts traditional knowledge practices that depend on specialized experts to disseminate knowledge through carefully regulated channels and institutions. Furthermore, social knowledge creation spaces like Wikipedia induce productive epistemic turbulence through multivocal authorship, arguments, and collaboration. Pfister champions this networked or participatory expertise as a more democratic, representative, and therefore less hierarchical model of communication.

Rosenzweig, Roy. 2006. "Can History Be Open Source? Wikipedia and the Future of the Past." *Journal of American History* 93 (1): 117–46.
Rosenzweig envisions a model for history scholarship based on the open access, multi-author Wikipedia framework. He concedes that Wikipedia represents an exciting—and perhaps even more ethical—structure of sharing and creating knowledge. Although Rosenzweig thoroughly and comprehensively acknowledges all of the criticisms of Wikipedia from an academic standpoint, he nonetheless proposes that history scholars become more open to incorporating Wikipedia in their scholarly practice. Rosenzweig heralds the many benefits of wiki-based learning and projects for both research and teaching purposes.

Siemens, Raymond G., Meagan Timney, Cara Leitch, Corina Koolen, and Alex Garnett, with the ETCL, INKE, and PKP Research Groups. 2012. "Toward Modeling the *Social* Edition: An Approach to Understanding the Electronic Scholarly Edition in the Context of New and Emerging Social Media." *Digital Scholarship in the Humanities* (formerly *Literary and Linguistic Computing*) 27 (4): 445–61. doi:10.1093/llc/fqs013.
Siemens, Timney, Leitch, Koolen, Garnett, et al. present a vision of an emerging manifestation of the scholarly digital edition: the social edition. The authors ruminate on both the potential and already-realized intersections between scholarly digital editing and social media. For Siemens et al., many scholarly digital editions do not readily employ the collaborative electronic

tools available for use in a scholarly context. The authors seek to remediate this lack of engagement, especially concerning opportunities to integrate collaborative annotation, user-derived content, folksonomy tagging, community bibliography, and text analysis capabilities within a digital edition. Furthermore, Siemens et al. envision the conceptual role of the editor—traditionally a single authoritative individual—as a reflection of facilitation rather than of didactic authority. A social edition predicated on these shifts and amendments would allow for increased social knowledge creation by a community of readers and scholars, academics and citizens alike.

Wasik, Bill. 2009. *And Then There's This: How Stories Live and Die in Viral Culture.* New York: Viking.
Wasik explores what is at stake with viral culture and social networking, and celebrates the community-generated culture the Internet provokes. He argues that the proliferation of short- lived sensations common to the Internet has altered the way contemporary society creates knowledge and culture. Wasik details his experiences as creator of the first flash mob in Manhattan in 2003. He also explores various memes or "nanostories"—brief moments of celebrity facilitated by digital culture. Wasik concludes by urging readers to process information responsibly in order to resist getting lost spiritually or creatively in the deluge of temporally minute gasps of popular culture.

5. Spatial Humanities and Digital Mapping

Ancient World Mapping Center, Stoa Consortium, and Institute for the Study of the Ancient World. 2000. *Pleiades.* http://pleiades.stoa. org/.
Pleiades is a community-built gazetteer and graph of ancient places; its platform integrates embedded, interactive maps and scholarship written on the gazetteer itself. The major focus of *Pleiades* is on the Greek and Roman world, but this has expanded to include Ancient Near Eastern, Byzantine, Celtic, and Early Medieval geography. It operates under an open licence, and allows users to create and share digital historical maps about the ancient world. Users can also track updates and modifications of the map and texts through *Pleiades'* interface.

Bachelard, Gaston. 1969. *The Poetics of Space.* Translated by Maria Jolas. Boston: Beacon Press.
Rather than focusing on actual geographical space, Bachelard adopts a phenomenological approach in studying interior, "poetic" spaces such as those of the imagination, the memory, and the domestic sphere and objects within

it. He focuses on the term "topophilia," which is used to describe spaces that evoke an emotional reaction and that often reappear in people's imagination and memories throughout their lives. From this stems his argument for prioritizing topoanalysis over psychoanalysis, since in the process of returning to these memories, the spatial aspect is evoked more strongly than a discrete temporal one. Significantly, Bachelard addresses the concept of the "poetic image"—the ability of readers to capture the image described by the author without having direct visual access to that place.

Gregory, Derek. 1994. *Geographical Imaginations.* **Oxford: Blackwell.**
Gregory explores how different theoretical aspects of geography relate to a wide range of disciplines. His main objective is to situate geography as a discourse within contemporary thought, rather than accepting its segregation as a separate discipline. Gregory presents numerous examples of the role of geography in theoretical discussions, including feminism, postcolonialism, and postmodernism, and suggests that geography could help redefine these discourses. According to Gregory, social life is deeply situated in different aspects of spatiality, both literal and metaphorical.

Guldi, Jo, and Cora Johnson-Roberson. 2012. *Paper Machines.* **metaLAB @ Harvard. http://papermachines.org/.**
Guldi and Johnson-Roberson demonstrate Paper Machines, a plug-in for the open bibliographic management system Zotero. Paper Machines has a number of embedded analytical digital tools, one of which is a digital map that automatically geoparses the selected texts onto a map interface. Zotero libraries can be shared or made public, allowing people to experiment with these tools on the same corpus. A significant feature of Paper Machines is that it exports geoparsed information into JSON or CSV format free of charge, allowing this data to be reused on other mapping platforms. Paper Machines is also well equipped for dealing with large data sets, which allows researchers to investigate substantial collections.

Jenstad, Janelle, and Kim McLean-Fiander. n.d. "The *MoEML* **Gazetteer of Early Modern London."** *The Map of Early Modern London,* **edited by Janelle Jenstad. Victoria: University of Victoria. http://mapoflondon.uvic.ca/gazetteer_about.htm.**
The *MoEML* Gazetteer of Early Modern London is a digitized gazetteer that offers a standard for London place names from 1550 to 1650. This gazetteer consists of six components: a variant toponym or place name, the authority name, an @xml:id to accurately manage the data, a link to the Agas map interface, other alternate names and spellings, and a location type. Users

can contribute to the gazetteer by suggesting information to add. It is also openly available for researchers to adapt to their own projects. Each place in the gazetteer is linked to an encyclopedia page where a detailed description of the place can be found.

Moretti, Franco. 1998. *Atlas of the European Novel, 1800–1900*. London: Verso.

Moretti proposes using maps as analytical tools for an in-depth analysis of literature by adopting the concept of mapping as a way of reading. He argues that this approach reveals the rich and multilayered nature of literary works that may otherwise elude the reader, providing evidence for this claim by presenting and analyzing numerous digital maps of well-known works. Moretti is critical, however, of the fact that literary analysis focuses primarily on canonical work. He suggests that the focus ought to be expanded to include the immense body of marginalized literature, and that working with such a large corpus would require collaborative digital research. According to Moretti, these types of digital mapping projects should actively constitute the literary field and form part of its discourse, rather than simply operate as another method for studying it.

OpenStreetMap Foundation. n.d. *OpenStreetMap*. https://www. openstreetmap.org.

OpenStreetMap is an editable map of the world that consists of a vast amount of location information, ranging from bus routes and bicycle trails to cafés and restaurants. It owes much of its success to its open access values, which circumvent the widespread commercialization of geospatial information. OpenStreetMap is a collaborative project with more than two million users who crowdsource data through a number of resources, such as GPS devices and aerial photography. The OpenStreetMap Foundation—whose mission is to provide an infrastructure for openly reusable digital geospatial information—supports this project.

Stanford Natural Language Processing Group. 2006. *Stanford Named Entity Recognizer (NER)*. http://nlp.stanford.edu/software/CRF-NER. html.

The Stanford Named Entity Recognizer (NER) is a Java Implementation that processes and labels words in a text, including names of a location, organization, or person. This allows for the automatic placename tagging of places with standardized spelling, which lies at the center of automatic digital spatial mapping practices. Made available by the Stanford Natural Language Processing Group, the Stanford NER tagger is open access, open source, and accessible and adaptable under a GNU General Public License. The Stanford NER tagger is often the starting point to redirect extracted data to a geoparser, which matches the placenames with their geographical coordinates. It has been expanded to work with languages other than English, such as Chinese, German, and Spanish, and has been optimized to work with a multitude of programming languages.

Tally, Robert T., Jr. 2013. *Spatiality*. London and New York: Routledge.
Tally engages in a thorough overview of existing scholarship in spatial theory that has been produced in the aftermath of the spatial turn. He argues for the importance of three main concepts in spatial humanities—literary cartography, literary geography, and geocriticism—that can provide a new and engaging way of looking at literary theory, criticism, and practice of spatiality. According to Tally, geocriticism should be seen as an interdisciplinary methodology whose main aim is to explore diverse cartographies critically and through different approaches, including mapping and spatial humanities, in order to keep pace with the rapidly shifting nature of spatial relations in a postmodern world.

Westphal, Bertrand. 2011. *Geocriticism: Real and Fictional Spaces*. Translated by Robert T. Tally, Jr. New York: Palgrave Macmillan.
Westphal introduces geocriticism as an interdisciplinary method in literary studies that encompasses the study of space from a multifocal, polysensoral, and intertextual perspective, with a stratigraphic vision in mind. He bases geocriticism on three main theoretical assumptions: spatiotemporality, transgressivity, and referentiality. Spatiotemporality is the aspect of the work delineating space-time. Transgressivity is described as recognition of the ever-shifting boundaries of real and fictional spaces, and referentiality as the relationship between the representation and the referent being in continuous oscillation or movement. Given the scope that Westphal envisions, and the attempt to include a large corpus of texts as a postmodern critique of grand narratives, the practice of geocriticism would have to be built on collaborative effort with reliance on technology. Westphal advocates geocritical analysis as a continuous exploration without a fixed end point, and as a new way of engaging with real and fictional spaces after the spatial turn.

Wick, Marc (founder), and Christophe Boutreux (developer). *GeoNames*. Männedorf, Switzerland: Unxos GmbH. http://www.geonames.org.
GeoNames, the largest open gazetteer, is a downloadable database that contains more than 10 million geographical names in a set of different

languages, as well as population and elevation information. It is equipped for dealing with ambiguities that arise when working with spatial data, such as allowing users to select from a list of possible options when a place name refers to more than one geographical entity and providing alternative spellings where appropriate. GeoNames is based on a user-friendly wiki platform that allows users to add and edit information. This database operates according to open access standards and exports its database daily to keep it updated.

Wrisley, David J., and the team at the American University of Beirut. 2016. *Linguistic Landscapes of Beirut* (formerly *Mapping Language Contact in Beirut*). http://llb.djwrisley.com/.

Linguistic Landscapes of Beirut is an example of a pedagogical implementation of socially created maps into the classroom environment. The focus is on data that reveals local features of language use. The main objective of this project is to capture instances of multilingualism in Arabic, English, and French based on cross-lingual wordplay, vernacular or language-specific use, and language mixing. The collected information and its metadata are stored on a map through mobile data collection using the Fulcrum application. Students update their data live, with an average of four additions per student per week over an entire semester. This project points to the abundance of local multilingual features expressed within the metropolitan area of Beirut.

6. Crowdsourcing

Carletti, Laura, Derek McAuley, Dominic Price, Gabriella Giannachi, and Steve Benford. 2013. "Digital Humanities and Crowdsourcing: An Exploration." *Museums and the Web 2013 Conference.* Portland: Museums and the Web LLC. http://mw2013.museumsandtheweb.com/paper/digital-humanities-and-crowdsourcing-an-exploration-4/.

Carletti, McAuley, Price, Giannachi, and Benford survey and identify emerging practices in current crowdsourcing projects in the digital humanities. They base their understanding of crowdsourcing on an earlier 2012 publication that defined crowdsourcing as an online, voluntary activity connecting individuals to an initiative via an open call. This definition was used to select the case studies for the current research. The researchers found two major trends in the 36 initiatives included in the study: crowdsourcing projects use the crowd to either (a) integrate/enrich/configure existing resources or (b) create/contribute new resources. Generally, crowdsourcing projects asked volunteers to contribute in terms of curating, revising, locating, sharing, documenting, or enriching materials. The 36 initiatives surveyed divided naturally into three categories in terms of project aims: public engagement, enriching resources, and building resources.

Causer, Tim, and Melissa Terras. 2014. "Crowdsourcing Bentham: Beyond the Traditional Boundaries of Academic History." *International Journal of Humanities and Arts Computing* 8 (1): 46–64. doi:10.3366/ ijhac.2014.0119.

Causer and Terras look back on some of the key discoveries that have been made from the *Transcribe Bentham* crowdsourced initiative. *Transcribe Bentham* was launched with the intention of demonstrating that crowdsourcing can be used successfully for both scholarly work and public engagement by allowing all types of participants to access and explore cultural material. Causer and Terras note that the majority of the work on *Transcribe Bentham* was undertaken by a small percentage of users, or "super transcribers." Only 15 per cent of the users have completed any transcription, and approximately 66 per cent of those users have transcribed only a single document—leaving a very select number of individuals responsible for the core of the project's production. Causer and Terras illustrate how some of the user transcription has contributed to our understanding of some of Jeremy Bentham's central values: animal rights, politics, and prison conditions. Overall, Causer and Terras demonstrate how scholarly transcription undertaken by a wide, online audience can uncover essential material.

Causer, Tim, Justin Tonra, and Valerie Wallace. 2012. "Transcription Maximized; Expense Minimized? Crowdsourcing and Editing *The Collected Works of Jeremy Bentham*." *Digital Scholarship in the Humanities* (formerly *Literary and Linguistic Computing*) 27 (2): 119– 37. doi:10.1093/llc/fqs004.

Causer, Tonra, and Wallace discuss the advantages and disadvantages of user-generated manuscript transcription using the *Transcribe Bentham* project as a case study. The intention of the project is to engage the public with the thoughts and works of Jeremy Bentham by creating a digital, searchable repository of his manuscript writings. Causer, Tonra, and Wallace preface this article by setting out five key factors the team hoped to assess in terms of the potential benefits of crowdsourcing: cost effectiveness, exploitation, quality control, sustainability, and success. Evidence from the project showcases the great potential for open access TEI-XML transcriptions in creating a long-term, sustainable archive. Additionally, users reported that they were motivated by a sense of contributing to a greater good and/or recognition. In the experience of *Transcribe Bentham*, crowdsourcing transcription may not have been the cheapest, quickest, or easiest route; the authors argue, however, that projects with a longer time frame may find this method both self-sufficient and cost-effective.

Causer, Tim, and Valerie Wallace. 2012. "Building a Volunteer Community: Results and Findings from *Transcribe Bentham*." *Digital Humanities Quarterly* 6 (2): n.p. http://digitalhumanities.org:8081/ dhq/vol/6/2/000125/000125.html.

Causer and Wallace reflect on the experience of generating users and materials for the crowdsourced *Transcribe Bentham* project. The purpose of the *Transcribe Bentham* project is to create an open source repository of Jeremy Bentham's papers that relies on volunteers transcribing the manuscripts. Causer and Wallace argue that crowdsourcing is a viable and effective strategy only if it is well facilitated and gathers a group of willing volunteers. They found that retaining users was just as integral to the success of the project as was recruiting. It was important, therefore, that they build a sense of community through outreach, social media, and reward systems. The number of active users involved in *Transcribe Bentham* was greatly affected by media publicity. Users reported that friendly competition motivated them to participate, but that an overall lack of time limited their contributions.

Fitzpatrick, Kathleen. 2012. "Beyond Metrics: Community Authorization and Open Peer Review." In *Debates in the Digital Humanities,* edited by Matthew K. Gold, 452–59. Minneapolis: University of Minnesota Press. http://dhdebates.gc.cuny.edu/debates/text/7.

Fitzpatrick calls for a reform of scholarly communication via open peer review. She argues that the Internet has provoked a conceptual shift wherein (textual) authority is no longer measured by a respected publisher's stamp; rather, she contends, authority is now located in the community. As concepts of authority change and evolve in the digital sphere, so should methods. Peer review should be opened to various scholars in a field, as well as to non-experts from other fields and citizen scholars. Fitzpatrick claims that this sort of crowdsourcing of peer review could more accurately represent scholarly and non-scholarly reaction, contribution, and understanding. Digital humanities and new media scholars already have the tools to measure digital engagement with a work; now, a better model of peer review should be implemented to take advantage of the myriad, social, networked ways scholarship is (or could be) produced.

Franklin, Michael J., Donald Kossman, Tim Kraska, Sukriti Ramesh, and Reynold Xin. 2011. "CrowdDB: Answering Queries with Crowdsourcing." In *Proceedings of the 2011 ACM SIGMOD International Conference on Management of Data (SIGMOD/PODS '11)*, 61–72. New York: ACM.
Franklin, Kossman, Kraska, Ramesh, and Xin discuss the importance of including human input in query processing systems due to their limitations in dealing with certain subjective tasks, which often result in inaccurate results. The authors propose using CrowdDB, a system which allows for crowdsourcing input when dealing with incomplete data and subjective comparison cases. They discuss the benefits and limitations of having human effort combined with machine processing, and offer a number of suggestions to optimize the workflow. The authors envision the field of human input combined with computer processing to be an area of rich research due to its improvement of existing models and enablement of new ones.

Ghosh, Arpita, Satyen Kale, and Preston McAfee. 2011. "Who Moderates the Moderators? Crowdsourcing Abuse Detection in User-Generated Content." In *Proceedings of the 12th ACM Conference on Electronic Commerce (EC '11)*, 167–76. New York: ACM.
Ghosh, Kale, and McAfee address the issue of how to moderate the ratings of users with unknown reliability. They propose an algorithm that can detect abusive content and spam, starting with approximately 50 per cent accuracy on the basis of one example of good content, and reaching complete accuracy after a number of entries based on machine-learning techniques. They believe that rating each individual contribution is a better approach than rating the users themselves based on their past behaviour, as most platforms do. According to the authors, this algorithm may be a stepping-stone in determining more complex ratings by users with unknown reliability.

Holley, Rose. 2010. "Crowdsourcing: How and Why Should Libraries Do It?" *D-Lib Magazine* 16 (3/4): n.p. doi:10.1045/march2010–holley.
Holley defines crowdsourcing, and makes a number of practical suggestions to assist with launching a crowdsourcing project. She asserts that crowdsourcing uses social engagement techniques to help a group of people work together on a shared, usually significant initiative. The fundamental principle of a crowdsourcing project is that it usually entails greater effort, time, and intellectual input than is available from a single individual, thereby requiring broader social engagement. Holley's argument is that libraries are already proficient at public engagement, but need to improve how they work toward shared group goals. She suggests ten basic practices

to assist libraries in successfully implementing crowdsourcing. Many of these recommendations centre on project transparency and motivating users.

*Kittur, Aniket, and Robert E. Kraut. 2008. "Harnessing the Wisdom of the Crowds in Wikipedia: Quality Through Coordination." In *Proceedings of the 2008 ACM Conference on Computer Supported Cooperative Work (CSCW 08)*, 37–46. New York: ACM.
Kittur and Kraut study the correlation between the number of editors on a Wikipedia page and the quality of that page's content. Significantly, they argue that an increased number of editors on a given page will prove productive only if some sort of coordination apparatus is in place. Articles are even more successful, content-wise, if a small group of experts manages the majority of the work. This argument runs counter to the crowdsourcing ethos of Wikipedia, which dictates that, generally, the more editors at work, the better the quality of the article. The authors argue, however, that a smaller group of editors working under a semi-authoritative organizational system facilitates peer-to-peer communication—a benefit that is often lost when large groups of uncoordinated individuals are involved.

Manzo, Christina, Geoff Kaufman, Sukdith Punjasthitkul, and Mary Flanagan. 2015. "'By the People, For the People': Assessing the Value of Crowdsourced, User-Generated Metadata." *Digital Humanities Quarterly* 9 (1): n.p. http://www.digitalhumanities.org/dhq/vol/9/1/000204/000204.html.
Manzo, Kaufman, Punjasthitkul, and Flanagan make a case for the usefulness of folksonomy tagging when combined with categorical tagging in crowdsourced projects. The authors open with a defence of categorization by arguing that classification systems reflect collection qualities while allowing for efficient retrieval of materials. However, they admit that these positive effects are often diminished by the use of folksonomy tagging, which promotes self-referential and personal task organizing labels. The authors suggest that a mixed system of folksonomic and controlled vocabularies be put into play in order to maximize the benefits of both approaches while minimizing their challenges. This is demonstrated through an empirical experiment in labelling images from the Leslie Jones Collection of the Boston Public Library, followed by evaluating the helpfulness of the tags using a revised version of the Voorbij and Kipp scale.

McKinley, Donelle. 2012. "Practical Management Strategies for Crowdsourcing in Libraries, Archives and Museums." Report for

the School of Information Management, Faculty of Commerce and Administration, Victoria University of Wellington (New Zealand): n.p.   http://nonprofitcrowd.org/wp-content/uploads/2014/11/McKinley-2012–Crowdsourcing-management-strategies.pdf.

McKinley reviews the literature and theory on crowdsourcing, and considers how it relates to the research initiatives of libraries, archives, and museums. She begins by claiming that burgeoning digital technologies have contributed to an increase in participatory culture. Furthermore, she argues that this is evinced by the growing number of libraries, archives, and museums using crowdsourcing. McKinley cites five different categories of crowdsourcing: collective intelligence, crowd creation, crowd voting, crowdfunding, and games. By way of conclusion, McKinley makes the following recommendations for crowdsourcing projects: (a) understand the context and convey the project's benefits; (b) choose an approach with clearly defined objectives; (c) identify the crowd and understand its motivations; (d) support participation; (e) evaluate implementation.

Moyle, Martin, Justin Tonra, and Valerie Wallace. 2011. "Manuscript Transcription by Crowdsourcing: Transcribe Bentham." *Liber Quarterly* 20 (3–4): 347–56. doi:10.18352/lq.7999.

Moyle, Tonra, and Wallace outline the objectives of the *Transcribe Bentham* project from its initial stages. *Transcribe Bentham* hopes to harness the power of crowdsourcing to develop an open source repository of Jeremy Bentham's manuscripts. Beyond digitizing and transcribing the manuscripts, the project aims to create a transcription interface, promote community volunteerism, and roll out a TEI transcription tool, among other things. The authors work through the design concept for the transcription interface and the TEI toolbar, both of which are meant to mask the complexity of the markup. The project team hopes that this initiative will stimulate further public engagement in scholarly archives and that it will introduce Bentham's work to new audiences.

*OpenStreetMap Foundation. n.d. *OpenStreetMap.* https://www.openstreetmap.org.

OpenStreetMap is an editable map of the world that consists of a vast amount of location information, ranging from bus routes and bicycle trails to cafés and restaurants. It owes much of its success to its open access values, which circumvent the widespread commercialization of geospatial information. OpenStreetMap is a collaborative project with more than two million users who crowdsource data through a number of resources, such as GPS devices and aerial photography. The OpenStreetMap Foundation—whose mission is to provide an infrastructure for openly reusable digital geospatial information—supports this project.

Ridge, Mia. 2013. "From Tagging to Theorizing: Deepening Engagement with Cultural Heritage through Crowdsourcing." *Curator: The Museum Journal* **56 (4): 435–50. doi:10.1111/cura.12046.**
Ridge examines how crowdsourcing projects have the potential to assist museums, libraries, and archives with the resource-intensive tasks of creating or improving content about collections. She argues that a well-designed crowdsourcing project aligns with the core values and missions of museums by helping to connect people with culture and history through meaningful activities. Ridge synthesizes several definitions of crowdsourcing to present an understanding of the term as a form of engagement in which individuals contribute toward a shared and significant goal through completing a series of small, manageable tasks. She points toward several examples of such projects to illustrate her definition. Ridge argues that scaffolding the project by setting up boundaries and clearly defining activities helps to increase user engagement by making participants feel comfortable completing the given tasks. She sees scaffolding as a key component of mounting a successful crowdsourcing project that offers truly deep and valuable engagement with cultural heritage.

Rockwell, Geoffrey. 2012. "Crowdsourcing the Humanities: Social Research and Collaboration." In *Collaborative Research in the Digital Humanities,* **edited by Marilyn Deegan and Willard McCarty, 135–54. Farnham, UK, and Burlington, VT: Ashgate.**
Rockwell demonstrates how crowdsourcing can facilitate collaboration by examining two humanities computing initiatives. He exposes the paradox of collaborative work in the humanities by summarizing the "lone ranger" past of the humanist scholar. Rockwell asserts that the digital humanities are, conversely, characterized by collaboration because they require a diverse range of skills. He views collaboration as an achievable value of digital humanities rather than a transcendent one. Case studies of the projects *Dictionary* and *Day in the Life of Digital Humanities* illustrate the limitations and promises of crowdsourcing in the humanities. Rockwell argues that the main challenge of collaboration is the organization of professional scholarship. Crowdsourcing projects provide structured ways to implement a social, counterculture research model that involves a larger community of individuals.

Ross, Stephen, Alex Christie, and Jentery Sayers. 2014. "Expert/ Crowdsourcing for the Linked Modernisms Project." *Scholarly and*

Research Communication 5 (4): n.p. http://src-online.ca/index.php/
src/article/viewFile/186/368.

Ross, Christie, and Sayers discuss the creation and evolution of the Social Sciences and Humanities Research Council (SSHRC)-funded *Linked Modernisms Project.* The authors demonstrate how the project negotiates the productive study of both individual works and the larger field of cultural modernism through the use of digital, visual, and networked methods. *Linked Modernisms* employs a four-tier information matrix to accumulate user-generated survey data about modernist materials. The authors argue that the resulting information allows serendipitous encounters with data, and emphasizes discoverability. *Linked Modernisms* is focused on developing modes of scholarly publication that line up with the dynamic nature of the data and comply with the principles of open access.

Saklofske, Jon, with the INKE Research Group. 2012. "Fluid Layering: Reimagining Digital Literary Archives Through Dynamic, User-Generated Content." *Scholarly and Research Communication* 3 (4): n.p. http://src-online.ca/index.php/src/article/viewFile/70/181.

Saklofske argues that while the majority of print and digital editions exist as isolated collections of information, changing practices in textual scholarship are moving toward a new model of production. He uses the example of *NewRadial*, a prototype information visualization application, to showcase the potential of a more active public archive. Specifically, Saklofske focuses on making room for user-generated data that transforms the edition from a static repository into a dynamic and co-developed space. He champions the argument that the digital archive should place user-generated content in a more prominent position through reimagining the archive as a site of critical engagement, dialogue, argument, commentary, and response. In closing, Saklofske poses five open-ended questions to the community at large as a way of kickstarting a conversation regarding the challenges of redesigning the digital archive.

Walsh, Brandon, Claire Maiers, Gwen Nally, Jeremy Boggs, and Praxis Program Team. 2014. "Crowdsourcing Individual Interpretations: Between Microtasking and Multitasking." *Digital Scholarship in the Humanities* (formerly *Literary and Linguistic Computing*) 29 (3): 379–86. doi:10.1093/llc/fqu030.

Walsh, Maiers, Nally, Boggs, et al. track the creation of Prism, an individual text markup tool developed by the Praxis Program at the University of Virginia. Prism was conceived in response to Jerome McGann's call for textual markup tools that foreground subjectivity as the tool illustrates how different groups of readers engage with a text. Prism is designed to assist with projects that blend two approaches to crowdsourcing: microtasking and macrotasking. A compelling quality of Prism is that it balances the constraint necessary for generating productive metadata with the flexibility necessary for facilitating social, negotiable interactions with the textual object. In that way, Prism is poised to redefine crowdsourcing in the digital humanities.

7. Discipline Formation in the Academic Context

Ball, John Clement. 2010. "Definite Article: Graduate Student Publishing, Pedagogy, and the Journal as Training Ground." *Canadian Literature* 204: 160–62.

Ball speaks from his position as editor of the journal *Studies in Canadian Literature* on the social and pedagogical role of journals in graduate training and, thus, discipline formation. He suggests that academics view themselves as a part of a three-way pedagogical continuum that includes journals and graduate students. Although journals should not replace supervisors, they can play a significant role in the professionalization of graduate students by reviewing, critiquing, and disseminating graduate work. In this way, graduate students are better prepared to face the post-convocation job market.

*Bazerman, Charles. 1991. "How Natural Philosophers Can Cooperate: The Literary Technology of Coordinated Investigation in Joseph Priestley's *History and Present State of Electricity* (1767)." In *Textual Dynamics of the Professions: Historical and Contemporary Studies of Writing in Professional Communities*, edited by Charles Bazerman and James Paradis, 13–44. Madison: University of Wisconsin Press.

Bazerman studies the role of early literature reviews through a thorough discussion and analysis of Joseph Priestley's *The History and Present State of Electricity* (1767). If, as Bazerman argues, literature reviews constitute potent sites of knowledge-sharing and dissemination in a community, then Priestley's volume represents the first literature review, since it details the history of electricity research and experiments. Priestley created a comprehensive, open-ended document that summarized the accepted state of the field as well as anomalies, discrepancies, and failures. Bazerman applauds Priestley for his active service in democratizing and disseminating knowledge.

Berkenkotter, Carol, Thomas N. Huckin, and John Ackerman. 1991. "Social Context and Socially Constructed Texts: The Initiation of a Graduate Student into a Writing Research Community." In *Textual*

Dynamics of the Professions: Historical and Contemporary Studies of Writing in Professional Communities, **edited by Charles Bazerman and James Paradis, 191–215. Madison: University of Wisconsin Press.**

Berkenkotter, Huckin, and Ackerman develop a case study of a first-year graduate student's writing experience in order to discuss discipline formation via introduction to a discourse community. They argue that every shift into a new discursive, professional, or scholarly community requires a learning and application of discipline-specific rhetorical structures. Perhaps predictably, the authors conclude that previous relevant experience in a field better prepares a graduate student for rhetorical success. Although this conclusion appears initially obvious, it is pertinent when one considers current conversations surrounding graduate training reform. Overall, the authors present a unique study of graduate training and discipline formation through the lens of writing and rhetoric practices.

Berry, David M. 2011. "The Computational Turn: Thinking About the Digital Humanities." *Culture Machine* **12: n.p. http://www. culturemachine.net/index.php/cm/article/view/440/470.**

Berry narrates the formation of post-secondary education, traced back to Immanuel Kant's notion of reason as the guiding force of the ideal university. Berry maintains that the digital should now be considered the unifying idea of the contemporary university. He argues that the disparate, multiple knowledges produced in the university can unify via digital practice and context; by taking up the digital as form and content for educational institutions, we can move toward a more networked and decentralized "digital intellect" (7). This new ethos need not rely on traditional academic ideals of learning an entire literary canon or memorizing multiple equations. The focus would thus shift from the individual student or researcher to the collective, from the sharply delineated university to the postdisciplinary university.

*Biagioli, Mario. 2002. "From Book Censorship to Academic Peer Review." *Emergences* **12 (1): 11–45. doi:10.1080/1045722022000003435.**

Biagioli details the historical and epistemological shifts that have led to the academic peer review system as it is now known. Contrary to its contemporary role, peer review began as an early modern disciplinary technique closely related to book censorship and required for social and scholarly certification of institutions and individuals alike. The rise of academic journals shifted this constrained and royally-mandated position; no longer a self-sustaining system of judgment and reputation dictated by a small group of identified and accredited professionals, (often blind) peer review now focuses on disseminating knowledge and scholarship to the wider community. Biagioli also states that journals have moved from officially representing specific academic institutions to being community owned and operated, as responsibilities, duties, and readership are now dispersed among a group of like-minded scholars.

Brant, Claire. 2011. "The Progress of Knowledge in the Regions of Air?: Divisions and Disciplines in Early Ballooning." *Eighteenth-Century Studies* 45 (1): 71–86. doi:10.1353/ecs.2011.0050.

Brant studies discipline formation through the development and reception of balloons in the eighteenth century. She argues that, contrary to standard narratives about scientific discoveries and technological advances, discipline formation is in fact unruly and disorderly. In the case of balloons, it was this very disorder that drew a substantial amount of criticism from the "more serious" scientific community. Chaotic development also led to various "Eureka!" moments, and a thorough consideration of the possibilities and limitations of flight.

Brooks, Kevin. 2002. "National Culture and the First-Year English Curriculum: A Historical Study of 'Composition' in Canadian Universities." *American Review of Canadian Studies* 32 (4): 673–94. doi:10.1080/02722010209481679.

Brooks details the institutional, political, and economic history of composition courses at Canadian university English departments. Although first-year composition is a prominent fixture at American universities, it is not typically taught at their Canadian counterparts, something that surprises U.S. academics. Brooks discusses this notable curricular absence in the context of larger mid-twentieth-century fears concerning Canadian national identity and anti-Americanism. He suggests that the tension between Canadian and American views on composition "can function as a barometer for understanding the relationship between national cultures and higher education" (689).

Buehl, Jonathan, Tamar Chute, and Anne Fields. 2012. "Training in the Archives: Archival Research as Professional Development." *College Composition and Communication* 64 (2): 274–305.

Buehl, Chute, and Field discuss the possibilities for graduate training via archival research. The authors suggest that archival research is an appropriate avenue for professionalization, since it trains students to think and research methodically as well as to practise information literacy and management skills. Furthermore, archival research provokes a more nuanced understanding of historiography, preservation, and research practices. Through a case study, the authors prove the efficacy and benefits of training humanities scholars through archival methods.

Carlton, Susan Brown. 1995. "Composition as a Postdisciplinary Formation." *Rhetoric Review* 14 (1): 78–87. doi:10.1080/07350199509389053.
Carlton focuses on a specific moment of disciplinary formation in the field of composition. She outlines the arguments for and against composition becoming formally and nationally established as an academic discipline. Although many abhor the tenure-based credential system implicit in contemporary academic discipline formation, others argue that composition will not be taken seriously as a field until it is legitimized as a discipline. Carlton concludes in favour of composition as a discipline, but with a caveat of maintaining an enlightened, "postdiscipline" attitude.

*Eagleton, Terry. 2010. "The Rise of English." In *The Norton Anthology of
   Theory and Criticism*, edited by Vincent B. Leitch, 2140–46. New York:
   W.W. Norton.
Eagleton argues that the development of English literature was an ideological strategy used, beginning in the mid-nineteenth century, as a form of suppression and control to educate lower classes only "enough" to keep them subservient. English literature, moreover, was actually scorned and primarily directed at women when first introduced as a field of university study. Eagleton concludes that literature "is an ideology" (2140) due to its historical role in social development and nation-building in England and elsewhere.

Fitzpatrick, Kathleen. 2012. "The Humanities, Done Digitally." In
   Debates in the Digital Humanities, edited by Matthew K. Gold, 12–15.
   Minneapolis: University of Minnesota Press. http://dhdebates.
   gc.cuny.edu/debates/text/30.
Fitzpatrick addresses the long-standing debate of what should be considered the digital humanities, and addresses how the term has evolved over time to shift the focus from resting chiefly on the digital aspect. She tackles the existing tension between those who argue for digital humanities as the creation of knowledge and tools through digitized means, and those who believe it should be expanded to include interpretation. Through a set of examples, Fitzpatrick points to how this has been a long-standing issue in other aspects of the humanities. She concludes that the most productive outcome of such an argument is to bridge the opposing positions by considering the digital humanities as inherently interdisciplinary. Fitzpatrick adds that a distinctive

aspect of the field is its investigation of the ways in which the digital changes traditional humanities research and scholarly communication.

*Fjällbrant, Nancy. 1997. "Scholarly Communication—Historical Development and New Possibilities." In *Proceedings of the IATUL Conferences.* West Lafayette, IN: Purdue University Libraries e-Pubs. http://docs. lib.purdue.edu/cgi/viewcontent.cgi?article=1389&context=iatul.

In order to study the widespread transition of scholarly communication from print to electronic formats, Fjällbrant details the history of the scientific journal. Academic journals had emerged in seventeenth-century Europe, and the first of these, the *Journal des Sçavans*, was published in 1665 in Paris. The first learned societies formed at this time—the Royal Society in London and the Académie des Sciences in Paris—were primarily concerned with the dissemination of knowledge, and the scholarly journal developed out of a desire by researchers to share their findings with others in a cooperative forum. Following the lead of the Royal Society, some of whose members had read the *Journal des Sçavans*, other societies established similar serial publications. Although there were other contemporaneous forms of scholarly communication, including the letter, the scientific book, the newspaper, and the cryptic anagram system, the journal emerged as a primary source of scholarly communication. It met the needs of various stakeholders: the general public, booksellers and publishers, libraries, authors who wished to make their work public and claim ownership, the scientific community invested in reading and applying other scientists' findings, and academic institutions that required metrics for evaluating faculty.

Garson, Marjorie. 2008. "ACUTE: The First Twenty-Five Years, 1957–1982." *English Studies in Canada* 34 (4): 21–43. https://ejournals. library.ualberta.ca/index.php/ESC/article/view/19771/15285.

In this reprint of a pamphlet originally published in 1982, Garson discusses the first two and a half decades of ACUTE, now known as ACCUTE: the Association of Canadian College and University Teachers in English. While Garson comprehensively details the history of the association—the first conference, the development of the member base, the initial aims—she simultaneously notes the political and economic status of post-secondary English departments in Canada. Needless to say, this status has been tenuous and fraught almost from the inception of humanities departments in Canada. Overall, Garson provides an informative view into how the study of English literature has developed institutionally and socially, as well as a more specific glimpse into the trajectory of one of the major learned societies in Canada.

Graff, Gerald. 2003. "Introduction: In the Dark All Eggheads Are Gray."
In *Clueless in Academe: How Schooling Obscures the Life of the Mind*,
1–16. New Haven, CT: Yale University Press.

Graff argues that contemporary mores in academia constrict, rather than foster, social knowledge creation. A false intellectual divide exists that is heavily predicated on purposeful incomprehensibility in academic writing and practice. For Graff, academics render their communication more obscure than necessary because of underlying anxieties concerning irrelevancy, or worse, so-called vulgarity. Graff argues that academics, and perhaps most especially teachers, must avoid the trap of pretentious hyperintellectual rhetoric in order to actually inspire knowledge and to work together with students in the realm of higher education.

_____ 1987. *Professing Literature: An Institutional History*. Chicago:
University of Chicago Press.

Graff thoroughly details the history of twentieth-century English literature studies in America. He argues that many of the issues in contemporary academia can be traced to an overall method of patterned isolationism in a department. Due to intellectual or discipline-based conflicts, various isolated fields of thought and practitioners prevail. Conflicts are neither acknowledged nor attended to, but rather overlooked by a general attitude of inclusion and comprehensiveness. As a result, divergent schools of thought never engage in conversation or debate, and all practitioners are endowed with silos in which they can effectively ignore their intellectual opponents. The self-perpetuating lack of interconnectedness and collaboration in English departments has negatively affected their overall scholarship and success. Furthermore, Graff contends that the conflict between schools of thought (classicism, New Criticism, critical theory, and now, perhaps, digital humanities) should be taught to students in order to contextualize and lend meaning to their literary education. Graff presents the above arguments as an introduction to a comprehensive historical explanation of how literary studies evolved as a discipline.

Hayles, N. Katherine. 2008. *Electronic Literature: New Horizons for the*
Literary*. Notre Dame, IN: University of Notre Dame Press.

Hayles surveys the field known as electronic literature. She suggests that while electronic literature acknowledges the expectations formed by the print medium, it also builds on and transforms them. In addition, electronic literature is informed by other traditions in contemporary digital culture, including computer games. In this way, electronic literature embodies a hybrid of various forms and traditions that may not usually fit together. Hayles outlines a wide variety of electronic literature examples, and comments that new approaches of analysis are required—in particular, the ability to "think digital" and to recognize the aspects of networked and programmable media that do not exist in print literature. In electronic literature, neither the body nor the machine should be given theoretical priority. Instead, Hayles argues for interconnections that "mediate between human and machine cognition" (x). She sees this intermediation as a more playful form of engaging with the complex mix of possibilities offered by contemporary electronic literature.

Ittersum, Martine J. van. 2011. "Knowledge Production in the Dutch Republic: The Household Academy of Hugo Grotius." *Journal of the History of Ideas* 72 (4): 523–44. doi:10.1353/jhi.2011.0033.

Ittersum approaches early modern knowledge production through the lens of a seventeenth- century Dutch scholar, Hugo Grotius, and his family. She argues that scholarly families prevailed as units of knowledge production or household academies in the early modern period. Household academies were built upon a familial infrastructure of research, support, editing, and promotion. Significantly, Ittersum asserts that Grotius's success, in particular, depended largely on the diligent writerly and readerly efforts of his family.

*Jagodzinski, Cecile M. 2008. "The University Press in North America: A Brief History." *Journal of Scholarly Publishing* 40 (1): 1–20. doi:10.1353/scp.0.0022.

Jagodzinski describes the history of the North American university press, beginning with the first presses at Cornell and Johns Hopkins Universities, which debuted in the nineteenth century. From the beginning, the primary function of the university press was considered to be the dissemination of knowledge. Twentieth-century growth in the number of colleges and universities led to a corresponding growth in the number of university presses, and the Association of American University Presses (AAUP) was formally established in the mid-1930s. As is well known, the last quarter of the twentieth century heralded major systemic changes and obstacles, and the university press was not immune to these challenges. Jagodzinski discusses in detail how university presses have responded, pragmatically and creatively, to the (largely financial) issues burdening contemporary scholarly communication.

Jones, Steven E. 2013a. *The Emergence of the Digital Humanities*. London and New York: Routledge.

Jones studies the emergence of digital humanities in response to changes in culture. He uses William Gibson's concept of the eversion of cyberspace (that is, the boundary crossing, flipping, and erasure between cyberspace and non-cyberspace) as a way to describe the cultural change that has led to the current incarnation of digital humanities. Jones frames the emergence of digital humanities as a blending of textual studies and game studies. He provides readings of popular games such as *Fez* and *Spore*, as well as a number of indie games, to analyze the relation between digital humanities and game studies. Jones concludes with an overview of practices (such as desktop fabrication) that are relevant to both gaming and digital humanities. For a snapshot of Jones's stated views on scholarly communication, please see the annotation of the "Publication" chapter from *The Emergence of the Digital Humanities* elsewhere in this bibliography collection.

Kaufer, David S., and Kathleen M. Carley. 1993. "Academia." In *Communication at a Distance: The Influence of Print on Sociocultural Organization and Change*, 341–93. Hillsdale, NJ: Lawrence Earlbaum Associates.

Through sociological methods, Kaufer and Carley explore the relationship between academia and print culture. The authors concede that shared, participatory textual conventions enforce stability in academic professions, as one of the significant and most obvious effects of print is to enhance the speed and efficiency of information travelling to and through communities. Kaufer and Carley run a set of simulations in order to explore the dissemination of ideas in an academic discipline; they concur that rapid advances, social knowledge creation, and a growing community all depend on the efficacy of print dissemination. As such, a disciplinary familiarity with the form and allowances of print proves desirable for academic writers. Notably, the authors briefly touch on the interrelations between the Royal Societies, scientific journals, and print.

Kirschenbaum, Matthew. 2012. "What is Digital Humanities and What's It Doing in English Departments?" In *Debates in the Digital Humanities*, edited by Matthew K. Gold, 3–11. Minneapolis: University of Minnesota Press. http://dhdebates.gc.cuny.edu/debates/text/38.

Kirschenbaum discusses the origins, evolution, and distinctive features of the digital humanities as a field. He draws attention to the strong feelings of community, common purposefulness, and valued openness in the digital humanities, which he believes can resist the present crisis the humanities face. Kirschenbaum defends the position of digital humanities in English departments, and portrays how digital humanities embraces modern scholarship. He also claims that digital humanities are the first "new big thing" in the humanities in a long time. For Kirschenbaum, digital humanities practices of scholarship and pedagogy are based on transparency, a deep sense of engagement, and a strong network of scholars and researchers.

Lightman, Harriet, and Ruth N. Reingold. 2005. "A Collaborative Model for Teaching E-Resources: Northwestern University's Graduate Training Day." *Libraries and the Academy* **5 (1): 23–32. doi:10.1353/ pla.2005.0008.**

Lightman and Reingold expand on the annual Graduate Training Day held by the library at Northwestern University. This program aims to increase the information literacy of incoming graduate students. Ideally, Graduate Training Day will better prepare students for their upcoming scholarly practices as well as their professional lives after graduate school. Lightman and Reingold argue that information literacy is necessary training for graduate students, as it introduces bibliographic, research, digital humanities, and project management tools students may not be familiar with prior to their graduate education. (At the time of writing, it is unclear whether Graduate Training Day continues.)

Lorimer, Rowland. 2013. "Libraries, Scholars, and Publishers in Digital Journal and Monograph Publishing." *Scholarly and Research Communication* **4 (1): n.p. http://src-online.ca/index.php/src/article/view-File/43/117.**

Lorimer briefly details the last forty years of scholarly publishing to explicate the current state of affairs. He asserts that a reorganization of the academic publishing infrastructure would greatly encourage forthright contributions to knowledge, especially concerning academic journals and monographs. The splitting of the university press from the university (except in name), coupled with funding cuts and consequent entrepreneurial publishing projects, has hampered the possibilities of academic publishing. If all of the actors of digital scholarly communication—libraries, librarians, scholars on editorial boards, technologically-inclined researchers, programmers, digital humanists, and publishing professionals—were brought together in an inclusive collaboration, digital technology could yield significant benefits for the future of scholarship and knowledge creation.

*Moretti, Franco. 1998. *Atlas of the European Novel, 1800–1900.* **London: Verso.**

Moretti proposes using maps as analytical tools for in-depth analysis of literature by adopting the concept of mapping as a way of reading. He argues that this approach reveals the rich and multilayered nature of literary works that may otherwise elude the reader, providing evidence for this claim by presenting and analyzing numerous digital maps of well-known works. Moretti is critical, however, of the fact that literary analysis focuses primarily on canonical work. He suggests that the focus ought to be expanded to include the immense body of marginalized literature, and that working with such a large corpus would require collaborative digital research. According to Moretti, these types of digital mapping projects should actively constitute the literary field and form part of its discourse, rather than simply acting as another method for studying it.

Nowviskie, Bethany. 2012a. "A Digital Boot Camp for Grad Students in the Humanities." *The Chronicle of Higher Education*. Last modified April 29, 2012. http://chronicle.com/article/A-Digital-Boot-Camp-for-Grad/131665.

Nowviskie discusses the Praxis program she directs out of the Scholars' Lab at the University of Virginia. She demonstrates how a combined commitment to interdisciplinarity, collaboration, and tacit knowledge is used to effectively train graduate students in contemporary humanities (and especially digital humanities) work. Nowviskie acknowledges the challenges and benefits of blending radically new methods for graduate training with traditional humanities practices and credit systems. Overall, she reiterates the value of training graduate students in an open-ended, community-minded way; in this way, humanities programs can facilitate both graduate and postgraduate school careers.

*Siemens, Raymond G. 2002. "Scholarly Publishing at its Source, and at Present." In *The Credibility of Electronic Publishing: A Report to the Humanities and Social Sciences Federation of Canada*, compiled by Raymond G. Siemens, Michael Best, Elizabeth Grove-White, Alan Burk, James Kerr, Andy Pope, Jean-Claude Guédon, Geoffrey Rockwell, and Lynne Siemens. *TEXT Technology* 11 (1): 1–128.

Siemens's introduction to this report focuses on the rethinking of scholarly communication practices in light of new digital forms. He meditates on this topic through the framework of *ad fontes*—the act, or conception, of going to the source. As he argues, scholars should look at the source or genesis of scholarly communication. For Siemens, the source goes beyond the seventeenth-century inception of the academic print journal to include less formal ways of communicating and disseminating knowledge—i.e., verbal exchanges, epistolary correspondence, and manuscript circulation. In this way, scholars can look past the popular, standard academic journal and into a future of scholarly communication that productively involves varied scholarly traditions and social knowledge practices.

Svensson, Patrik. 2012. "Beyond the Big Tent." In *Debates in the Digital Humanities,* edited by Matthew K. Gold, 36–49. Minneapolis: University of Minnesota Press. http://dhdebates.gc.cuny.edu/debates/text/22.

Svensson explores the all-inclusive "big tent" notion of digital humanities by investigating its applicability in reality through concrete examples. He demonstrates that it is not always the case that digital humanities, as a field, is open to all. In some cases, the field may actually be disadvantageous to the people working within it. He argues that digital humanities should be seen instead as a trading zone and meeting place as it focuses more on interdisciplinary work and points to the intersection with many traditional disciplines. An emphasis is placed on the advantage of digital humanities remaining in a liminal position rather than becoming a discipline, since this makes it more inclusive and allows the digital humanities to play a positive role in shaping the humanities as a whole.

*Westphal, Bertrand. 2011. *Geocriticism: Real and Fictional Spaces.* Translated by Robert T. Tally, Jr. New York: Palgrave Macmillan.

Westphal introduces geocriticism as an interdisciplinary method in literary studies that encompasses the study of space from a multifocal, polysensoral, and intertextual perspective, with a stratigraphic vision in mind. He bases geocriticism on three main theoretical assumptions: spatiotemporality, transgressivity, and referentiality. Spatiotemporality is the aspect of the work delineating space-time. Transgressivity is described as recognition of the ever-shifting boundaries of real and fictional spaces, and referentiality as the relationship between the representation and the referent being in continuous oscillation or movement. Given the scope that Westphal envisions, and the attempt to include a large corpus of texts as a postmodern critique of grand narratives, the practice of geocriticism would have to be built on collaborative effort with reliance on technology. Westphal advocates geocritical analysis as a continuous exploration without a fixed end point, and as a new way of engaging with real and fictional spaces after the spatial turn.

Zacharias, Robert. 2011. "The Death of the Graduate Student (and the Birth of the HQP)." *English Studies in Canada* 37 (1): 4–8. https://ejournals.library.ualberta.ca/index.php/ESC/article/view/25197/18694.

Zacharias calls for increased attention to the changing role of humanities graduate students to that of "highly qualified personnel" (HQP). For the author, the shift represents a widespread aversion toward the humanities, and graduate studies (and students) in particular. Zacharias suggests that this reconsideration (and, in his view, corporatization) of graduate students be quelled, and that graduate education be considered just that: education, not training. He advocates for a more effective systematic introduction to the academic field by refocusing on comprehensive mentorship and humanities-based professionalization.

8. Public Humanities

Avila, Maria, with contributions from Alan Knoerr, Nik Orlando, and Celestina Castillo. 2010. "Community Organizing Practices in Academia: A Model, and Stories of Partnerships." *Journal of Higher Education Outreach and Engagement* **14 (2): 37–63. http://openjournals. libs.uga.edu/index.php/jheoe/article/view/43/38.**
Avila shares the details of her model of civic engagement at Occidental College in Northeast Los Angeles—a model focused on practical long-term reciprocal partnerships between communities and academics, rather than on abstracted discourse about the issues involved in maintaining these partnerships. Avila's model includes assessing the interest of college members (e.g., faculty, community partners, and students), building a leadership team, creating dynamic strategies and programs, and engaging in critical reflection. She concludes by speculating whether other institutions eager to build academic community partnerships in order to bring about positive cultural and social change could adopt her model.

Brown, David W. 1995. "The Public/Academic Disconnect." In *Higher Education Exchange Annual,* **38–42. Dayton, OH: Kettering Foundation.**
Brown suggests that cuts in financial support—as a result of the unwillingness of the public to support institutions—are at the root of the crisis in higher education. He argues that this is a natural reaction, given that many colleges and universities are often disengaged from the interests of the public and may not benefit the community in obvious ways. According to Brown, this calls for a rethinking of institutional practices, and could be remedied by having academics engage in public problem-solving *with* the community, rather than by merely talking *to* the community. Brown argues that universities or regional consortiums should establish civic training centres in which academics and members of the community can engage in productive discourse toward problem-solving strategies together. In these training centres, faculty members could also offer students the necessary skills to approach situations that resemble real-life complexities and diversities. Brown believes that if an entire diverse campus worked actively toward a public goal, the rhetoric of multiculturalism could move closer to being realized in practice.

Ellison, Julie. 2008. "The Humanities and the Public Soul." *Antipode* **40 (3): 463–71. doi:10.1111/j.1467-8330.2008.00615.x.**

Ellison describes public scholarship in the arts and humanities as having the distinct quality of an educated hopefulness. The unique discourse of public scholarship can confront complex issues, serve as an opportunity for practical experimentation, and participate publicly for the benefit of the larger community. Ellison argues that there are several models available to link public scholarship discourse to economies of cultural work in order to set it into practice. For Ellison, a space must be created for these discourses to be held and expanded by representatives of the university and the community. She provides as an example the Imagining America consortium, which consists of individuals, institutions, and associations whose agenda is to move public interest to the centre of higher education, often through multidisciplinary project-based work. Ellison envisions the future of public scholarship in the humanities and arts to be a space where discouragement, uncertainties, and complexities voiced by community partnerships would be seen as conditions of possibility and opportunities for the production of new knowledge.

Ellison, Julie, and Timothy Eatman. 2008. *Scholarship in Public: Knowledge Creation and Tenure Policy in the Engaged University.* **Syracuse, NY: Imagining America. http://imaginingamerica.org/wp-content/ uploads/2011/05/TTI_FINAL.pdf**

Ellison and Eatman discuss how the administrative side of American universities lags behind in terms of tenure and promotion policies, despite the increase in publicly engaged academic work. Through a series of interviews and substantial research, Ellison and Eatman clearly outline the current position of scholars doing work in publicly engaged fields and the anxieties involved in pursuing this work, including the strong discouragement by many universities themselves. The report serves as a guide for members of the university to change the position of publicly engaged scholarship so as to make it appropriate for career development by adapting policies regarding tenure and promotion. Ellison and Eatman address the importance of adjusting university policies according to informed graduate student demands, so that publicly engaged scholars of the future will stay and thrive on campus. The authors also acknowledge that adjusting policies is only part of the process; they offer a pathway to a larger change with regard to present conceptions of "peer" and "publication" that would make the production of knowledge more inclusive on campus and in the community.

Farland, Maria M. 1996. "Academic Professionalism and the New Public Mindedness." *Higher Education Exchange* Annual: 51–57. http://www.unz.org/Pub/HigherEdExchange-1996q1-00051.

Farland points to the sudden increase of scholars interested in the public, especially following the recent discontent expressed against universities for their relative absence from the public sphere. However, examining this interest more closely, Farland argues that academics have picked up the term "public" merely as an opportunity to develop their careers in their specialized disciplinary domains rather than to address the actual needs of wider society as demanded by politicians, media, and the public at large. This can be remedied by engaging higher education in public problem-solving and debates. Farland concludes that the new public mindedness will allow the university to maintain its present status, and to restore its relation to public life, when academic practices are brought into a direct conversation with the problems a community faces.

Haft, Jamie. 2012. "Publicly Engaged Scholarship in the Humanities, Arts, and Design." *Animating Democracy*: 1–15. http://imaginingamerica.org/wp-content/uploads/2015/09/JHaft-Trend-Paper.pdf.

Haft points to exemplary projects and initiatives in American campus-community practices through the lens of Imagining America: Artists and Scholars in Public Life, a "national coalition working explicitly at the nexus of publicly engaged scholarship and the humanities, arts, and design" (2). Her main argument is for the urgency of building, maintaining, and expanding an infrastructure for publicly engaged scholarship. This infrastructure, Haft believes, could solve pressing issues from a local to a national scale. She points to several existing practices in publicly engaged scholarship that share a common set of values, including community/campus engagement and dialogue, improving communities though collaboratively produced knowledge, and assuring cultural diversity and social equity. Haft concludes by arguing that the humanities should not be dismissed in favour of the sciences, because the humanities offer efficient ways to address and deal with urgent problems in their complexities.

Jay, Gregory. 2012. "The Engaged Humanities: Principles and Practices for Public Scholarship and Teaching." *Journal of Community Engagement and Scholarship* 3 (1): 51–63. http://jces.ua.edu/the-engaged-humanities-principles-and-practices-for-public-scholarship-and-teaching/.

Jay addresses the new role that the humanities can adopt in face of the present funding crisis in higher education due to its perceived absence from the public sphere. He argues that the humanities should work toward implementing a problem-based learning approach that equips students and faculty to deal with complex public issues more efficiently. Jay also advocates for the continued shift toward a digital mode of learning and scholarship. Through clear examples and guidelines, Jay outlines the concrete steps that should be taken at the institutional level to ensure that this shift is meaningful for the members of both the university and the community. He addresses the vital role that new media, the Internet, and technology play in higher education, and its relation to the public sphere, especially in terms of the production and dissemination of knowledge to a wider audience. Although it is still unclear how to assess the scope of the humanities' benefit for the public sphere, Jay believes that if the campus were an active member of the community, rather than a passive observer, this would be a step in the right direction.

9. The Shifting Future of Scholarly Communication and Digital Scholarship

Arbuckle, Alyssa, Constance Crompton, and Aaron Mauro. 2014. Introduction: "Building Partnerships to Transform Scholarly Publishing." *Scholarly and Research Communication* **5 (4): n.p. http://src-online.ca/index.php/src/article/view/195.**
Arbuckle, Crompton, and Mauro introduce the proceedings for an Implementing New Knowledge Environments (INKE) gathering in Whistler, Canada, in February 2014. This introduction reflects a network of scholarly communication players and possibilities in Canada. The editors acknowledge the pressing areas of concern around scholarly communication for the authors in the collection, including the fluidity of digital publication, the opportunity for (and challenges of) open peer review, and the effects of rapid technological development on scholarship. Arbuckle, Crompton, and Mauro enumerate suggestions made at the gathering proper, and suggest ways forward for the multi-disciplinary group that came together for this event.

Arbuckle, Alyssa, Aaron Mauro, and Lynne Siemens. 2015. Introduction: "From Technical Standards to Research Communities: Implementing New Knowledge Environments Gatherings, Sydney 2014 and Whistler 2015." *Scholarly and Research Communication* **6 (2): n.p. http://src-online.ca/index.php/src/article/view/232.**
Arbuckle, Mauro, and Siemens introduce the proceedings for two Implementing New Knowledge Environments (INKE) gatherings, in Sydney, Australia (December 2014) and Whistler, Canada (January 2015). The editors ruminate on the closing gap between scholarly communication and digital

practices, and draw together groupings of papers on relevant topics including collaboration, networked scholarship, knowledge production, prototyping, and pragmatic challenges. Arbuckle, Mauro, and Siemens also detail the alternative peer review method employed for the proceedings. Overall, the editors consider the breadth and depth of current scholarly communication and digital scholarship activities.

*Berry, David M. 2011. "The Computational Turn: Thinking About the Digital Humanities." *Culture Machine* 12: n.p. http://www. culturemachine.net/index.php/cm/article/view/440/470.

Berry narrates the formation of post-secondary education, traced back to Immanuel Kant's notion of reason as the guiding force of the ideal university. Berry maintains that the digital should now be considered the unifying idea of the contemporary university. He argues that the disparate, multiple knowledges produced in the university can unify via digital practice and context; by taking up the digital as form and content for educational institutions, we can move toward a more networked and decentralized "digital intellect" (7). This new ethos need not rely on traditional academic ideals of learning an entire literary canon or memorizing multiple equations. The focus would thus shift from the individual student or researcher to the collective, from the sharply delineated university to the postdisciplinary university.

Besser, Howard. 2004. "The Past, Present, and Future of Digital Libraries." In *A Companion to Digital Humanities*, edited by Susan Schreibman, Raymond G. Siemens, and John Unsworth, 557–75. Oxford: Blackwell.

Besser examines the state and trajectory of digital libraries, and argues that further considerations must be made in order for digital libraries to uphold both the tenets and roles of traditional libraries. After briefly surveying the position, history, and standards of the library, Besser concludes that libraries have certain key components that must be acknowledged and upheld in digital substantiations. These components include interoperability, stewardship, service, privacy, and equal access to a diversity of information. Besser argues that these components reflect the ethical side of the library, and need to be considered alongside more obvious priorities of information dissemination and preservation of artifacts.

*Biagioli, Mario. 2002. "From Book Censorship to Academic Peer Review." *Emergences* 12 (1): 11–45. doi:10.1080/1045722022000003435.

Biagioli details the historical and epistemological shifts that have led to the academic peer review system as it is now known. Contrary to its contemporary role, peer review began as an early modern disciplinary technique closely related to book censorship and required for social and scholarly certification of institutions and individuals alike. The rise of academic journals shifted this constrained and royally-mandated position; no longer a self-sustaining system of judgment and reputation dictated by a small group of identified and accredited professionals, (often blind) peer review now focuses on disseminating knowledge and scholarship to the wider community. Biagioli also states that journals have moved from officially representing specific academic institutions to being community owned and operated, as responsibilities, duties, and readership are now dispersed among a group of like-minded scholars.

Borgman, Christine L. 2007. *Scholarship in the Digital Age: Information, Infrastructure, and the Internet.* Cambridge, MA: MIT Press.
Borgman lays out research questions and hypotheses concerning the evolving scholarly infrastructure and modes of communication in the digital environment. She deduces that the inherent social elements of scholarship endure, despite new technologies that alter significantly the way scholarship is performed, disseminated, and archived. Scholarship and scholarly activities continue to exist in a social network of varying actors and priorities. Notably, Borgman focuses on the "data deluge"—the increasing amount of generated data and data accessed for research purposes. The influences of large data sets, as well as how these data sets will be preserved in keeping with library and archival conventions, are subjects of particular significance. Overall, Borgman synthesizes the various aspects of contemporary scholarship, and reflects on the increasingly pervasive digital environment.

Bowen, William R., Constance Crompton, and Matthew Hiebert. 2014. "Iter Community Prototyping an Environment for Social Knowledge Creation and Communication." *Scholarly and Research Communication.* 5 (4): n.p. http://src-online.ca/index.php/src/article/view/193/360.
Bowen, Crompton, and Hiebert discuss the features and challenges of Iter Community, a collaborative research environment. They also discuss *A Social Edition of the* Devonshire Manuscript, especially focusing on its human and computer social engagement. The authors organize the article into three sections: (1) a historical and conceptual framework of Iter Community; (2) an update on the state of Iter Community (at writing), and (3) a perspective on *A Social Edition of the* Devonshire Manuscript. They conclude that Iter Community's vision is to provide a flexible environment for communication, exchange, and collaboration, which will evolve with its participants' priorities and challenges.

*Burdick, Anne, Johanna Drucker, Peter Lunenfeld, Todd Presner, and Jeffrey Schnapp. 2012. "The Social Life of the Digital Humanities." In *Digital_Humanities*, 73–98. Cambridge, MA: MIT Press.

Burdick et al. focus on the social aspects and impacts of digital humanities. The authors argue that the digital humanities, by nature, encompass academic and social spaces that discuss issues beyond technology alone. Key issues include open access, open source publications, the emergence of participatory Web and social media technologies, collaborative authorship, crowdsourcing, knowledge creation, influence, authorization, and dissemination. Burdick et al. also consider the role of digital humanities in public spaces, beyond the siloed academy. The authors address these expansive issues through an oscillating approach of explanation and questioning. While the diversity of the topics in this chapter is substantial, the authors knit the arguments together under the broad theme of social engagement.

*Cohen, Daniel J. 2008. "Creating Scholarly Tools and Resources for the Digital Ecosystem: Building Connections in the Zotero Project." *First Monday* 13 (8): n.p. doi:10.5210/fm.v13i8.2233.

Cohen details how the Zotero project exemplifies both a Web 2.0 and a traditional scholarly ethos. He conceptualizes Zotero as a node in an interconnected digital ecosystem that builds bridges instead of hoarding information. Zotero is a widely used, open source, community-based bibliography tool. It exists on top of the browser as an extension, has maintained an API since its inception, and boasts comprehensive user features. As an easy-to-use collaborative tool, Zotero acts as both an effective scholarly resource and a facilitator of social knowledge creation.

Cohen, Daniel J., and Tom Scheinfeldt. 2013. "Preface." In *Hacking the Academy: New Approaches to Scholarship and Teaching from Digital Humanities*, edited by Daniel J. Cohen and Tom Scheinfeldt, 3–5. Ann Arbor: University of Michigan Press. doi:10.3998/dh.12172434.0001.001.

Cohen and Scheinfeldt introduce *Hacking the Academy*, a digital publishing experiment and attempt to reform academic institutions and practices by crowdsourcing content. The editors called for submissions to their project with the caveat that participants had one week to submit. Cohen and Scheinfeldt pitched their project with the following questions: "Can an algorithm edit a journal? Can a library exist without books? Can students build and manage their own learning management platforms? Can a conference be held without a program? Can Twitter replace a scholarly society?" (3). Roughly one sixth of the 329 submissions received were included in the consequent publication. The intent of the project was to reveal the desire and possibility for large institutional change via digital means.

Davidson, Cathy N., and David Theo Goldberg. 2004. "Engaging the Humanities." *Profession:* **42–62. doi:10.1632/074069504X26386.**
Davidson and Goldberg argue that despite marginalization, humanistic approaches and perspectives remain significant for successful, holistic university environments. Rather than taking a field-specific approach, Davidson and Goldberg propose a problem- or issue-based humanities model that allows for a more interdisciplinary approach. In this way, the comprehensive interpretive tools and complex models of cultural interaction integral to humanities work may resolve varied and continuous issues. The authors suggest that a conceptual and physical shift toward interdisciplinarities within institutions (rather than interdisciplinary institutions, models, or methods) offers a realistic and flexible approach to transforming academia and education.

De Roure, David. 2014. "The Future of Scholarly Communications." *Insights* **27 (3): 233–38. doi:10.1629/2048–7754.171.**
De Roure discusses how social machines provide a lens into the future developments in scholarship and scholarly collaboration. He considers the enhancement of the article as a mode of scholarly discourse, in ways that do not restrict innovation. De Roure examines scholarship at scale, as well as shifts in scholarship due to digital research and societal engagement. He continues to examine research objects and social machines in order to understand the evolution of digital scholarship. De Roure concludes that the article, the monograph, and the book need to be defamiliarized, and that the focus should shift to future practice.

Fitzpatrick, Kathleen. 2012. "Beyond Metrics: Community Authorization and Open Peer Review." In *Debates in the Digital Humanities,* **edited by Matthew K. Gold, 452–59. Minneapolis: University of Minnesota Press. http://dhdebates.gc.cuny.edu/debates/text/7.**
Fitzpatrick calls for a reform of scholarly communication via open peer review. She argues that the Internet has provoked a conceptual shift wherein (textual) authority is no longer measured by a respected publisher's stamp; rather, she contends, the community now locates authority. As concepts of authority change and evolve in the digital sphere, so should methods. Peer review should be opened to various scholars in a field as well as to non-experts from other fields and citizen scholars. Fitzpatrick claims that this sort of crowdsourcing of peer review could more accurately represent

scholarly and non-scholarly reaction, contribution, and understanding. Digital humanities and new media scholars already have the tools to measure digital engagement with a work; now, a better model of peer review should be implemented to take advantage of the myriad, social, networked ways scholarship is (or could be) produced.

_____ 2009. "Peer-To-Peer Review and the Future of Scholarly Authority." *Cinema Journal* 48 (2): 124–29. doi:10.1353/cj.0.0095.
Fitzpatrick explains that decentralized and displaced authority structures are taking over scholarly communication, and intellectual authority is shifting to spaces such as Wikipedia. Scholars must therefore embrace similarly open structures and public access, or the academic world will appear divorced from real world practices. For this reason, online peer-reviewed journals should not follow print practices of peer review, but must adapt and shape a new scholarly system. Current peer review processes do not necessarily ensure that the best work is in circulation, and in fact tend to replicate hierarchical and status-based privileges. Fitzpatrick argues for open process, web native modes of peer review in a peer-to-peer structure. Finally, she advocates for the need to articulate these values and standards to credentialing bodies in order for a more appropriate model of intellectual authorization to emerge.

_____ 2011. *Planned Obsolescence: Publishing, Technology, and the Future of the Academy.* New York: New York University Press. ("Introduction: Obsolescence," 1–14, and Chapter 3: "Texts," 89–120, are accessible at http://raley.english.ucsb.edu/wp-content2/uploads/234/Fitzpatrick.pdf.)
Fitzpatrick duly surveys and calls for a reform of academic publishing. She argues for more interactivity, communication, and peer-to-peer review, as well as a significant move toward digital scholarly publishing. Fitzpatrick demonstrates that the current mode of scholarly publishing is economically unviable. Moreover, tenure and promotion practices based primarily on traditional modes of scholarly publishing need to be reformed. Fitzpatrick acknowledges certain touchstones of the academy (peer review, scholarship, sharing ideas), and how these tenets have been overshadowed by priorities shaped, in part, by mainstream academic publishing practices and concepts. She details her own work with CommentPress, and the benefits of publishing online with an infrastructure that enables widespread dissemination as well as concurrent reader participation via open peer review.

*Fjällbrant, Nancy. 1997. "Scholarly Communication—Historical Development and New Possibilities." In *Proceedings of the IATUL Conferences.*

West Lafayette, IN: Purdue University Libraries e-Pubs. http://docs. lib.purdue.edu/cgi/viewcontent.cgi?article=1389&context=iatul.

In order to study the widespread transition of scholarly communication from print to electronic formats, Fjällbrant details the history of the scientific journal. Academic journals had emerged in seventeenth-century Europe, and the first of these, the *Journal des Sçavans*, was published in 1665 in Paris. The first learned societies formed at this time—the Royal Society in London and the Académie des Sciences in Paris—were primarily concerned with the dissemination of knowledge, and the scholarly journal developed out of a desire by researchers to share their findings with others in a cooperative forum. Following the lead of the Royal Society, some of whose members had read the *Journal des Sçavans*, other societies established similar serial publications. Although there were other contemporaneous forms of scholarly communication, including the letter, the scientific book, the newspaper, and the cryptic anagram system, the journal emerged as a primary source of scholarly communication. It met the needs of various stakeholders: the general public, booksellers and publishers, libraries, authors who wished to make their work public and claim ownership, the scientific community invested in reading and applying other scientists' findings, and academic institutions that required metrics for evaluating faculty.

*Flanders, Julia. 2009. "The Productive Unease of 21st-Century Digital Scholarship." *Digital Humanities Quarterly* 3 (3): n.p. http://www. digitalhumanities.org/dhq/vol/3/3/000055/000055.html.

Flanders discusses the role of the digital humanities in relation to the more conventional humanities, and characterizes the digital humanities as possessing a sort of "productive unease": anxiety concerning medium, institutional structures of scholarly communication, and representation. This anxiety is productive insofar as it brings into clearer focus previously unremarked upon biases in the traditional humanities. Moreover, digital tools and practices present more and different challenges. Of note, Flanders recognizes social software and media as tackling some of these anxiety-provoking issues, and acknowledges digital humanities projects that also strive in the same direction.

Guédon, Jean-Claude. 2008. "Digitizing and the Meaning of Knowledge." *Academic Matters* (October–November): 23–26. http://www.aca-demicmatters.ca/assets/AM_SEPT'08.pdf.

Guédon briefly sketches the recent history of scholarly communication and publishing and meditates on alternatives to the current state of affairs. He concludes that although open source publishing is a relatively recent phenomenon, it adroitly embodies the ethos and traditional practices of scholarship (especially in the sciences). For Guédon, open source publishing represents the open, endless appropriation of knowledge and discipline-wide conversation that has traditionally defined academic work. Guédon champions this move toward open, shared knowledge versus the continued exploitation of academics, librarians, and universities by the large corporate publishing companies currently relied upon for scholarly communication and accreditation.

Jones, Steven E. 2013. "Publications." In *The Emergence of the Digital Humanities*, 147–77. London and New York: Routledge.
Jones explores the current state of scholarly publishing and the role of the digital humanities. He argues that now, more than ever, academic practitioners are able to take the means of producing scholarly work into their own hands. Rather than relying on scholarly communication systems already in place, researchers can now experiment with different modes, media, and models of publication. Jones considers digital publishing and engagement of academic work as symptomatic of the deep integration and interplay of computational methods with contemporary scholarship in general, and digital humanities in particular.

Kingsley, Danny. 2013. "Build It and They Will Come? Support for Open Access in Australia." *Scholarly and Research Communication* 4 (1): n.p. http://src-online.ca/index.php/src/article/viewFile/39/121.
Kingsley surveys the longstanding, government-supported efforts toward open access scholarship that Australia has made. He focuses on mandates, publications, and repositories, and outlines both areas of success and potential improvement. Kingsley concedes that despite Australia's historic and widespread institutional and infrastructural commitment to open access, the country has not experienced a markedly higher uptake of open access production than the rest of the world. He points to the difficulty of clarifying and enforcing open access mandates, the inconsistency in funding operations and reporting, and the lack of an open access advocacy body as barriers to progressive movement in this area.

Lane, Richard J. 2014. "Innovation through Tradition: New Scholarly Publishing Applications Modelled on Faith-Based Electronic Publishing and Learning Environments." *Scholarly and Research Communication* 5 (4): n.p. http://src-online.ca/index.php/src/article/view/188.
Lane explores the popular eTheology platforms Olive Tree and Logos, and the possibilities for uptake of their information management and design models.

Lane details the advantages of popular or non-academic digital knowledge spaces and argues for their potential application to secular electronic publishing. The most evident advantage of this proposal may be the suggestion to tailor applications to communities of users, in the way that Olive Tree and Logos do, in order to develop a more integrated and dynamically engaged scholarly publishing system that includes user analysis.

*Liu, Alan. 2004. *The Laws of Cool: Knowledge Work and the Culture of Information*. Chicago: University of Chicago Press.

Liu interweaves two distinct threads in *The Laws of Cool*. He traces the history and ethos of "cool" (culture, trends, popularity, etc.) as well as postindustrial cool: the flux of cool knowledge work. Liu examines how the humanities can contribute to and survive in the new postindustrial cool, corporate landscape. Liu's sources and interests are widespread; he cites modernist design theory, Lev Manovich's database narrative, and everything from the Gayaki tribe to William Gibson's *Agrippa*. He concludes that the humanities are necessary to keep the corporation humane and informed of the history of its own practices; the humanities, in turn, must learn to negotiate the current cool cultural climate in order to remain relevant and effective.

*_____ 2012. "Where is Cultural Criticism in the Digital Humanities?" In *Debates in the Digital Humanities*, edited by Matthew K. Gold, 490–510. Minneapolis: University of Minnesota Press. http://dhdebates.gc.cuny.edu/debates/text/20.

Liu surveys the state of the digital humanities in relation to the humanities at large. He argues that, thus far, digital humanities projects have often lacked the self-reflexivity and cultural criticism necessary for the ethical development of humanistic projects—thereby denying the digital humanities a real or full position in the humanities. Because the digital humanities avoid cultural criticism, they frequently become subservient or merely instrumental to the humanities as a whole, functioning as either a moneymaker or tech support. Liu claims that the digital humanities could deconstruct the hierarchy by becoming both self-reflexive and invaluable, thereby leading the humanities into the academic future.

Lorimer, Rowland. 2014. "A Good Idea, a Difficult Reality: Toward a Publisher/Library Open Access Partnership." *Scholarly and Research Communication* 5 (4): n.p. http://src-online.ca/index.php/src/article/view/180.

Lorimer comments on the state of scholarly publishing in Canada. He offers thorough insight into the financial, social, and cultural obstacles that arise as academic institutions move toward an open access model of knowledge mobilization. Lorimer argues that although the *idea* of open access is desirable to academic and academic-aligned researchers, practitioners, and organizations, the *reality* of a complete open access model still requires considerable planning and implementation. Lorimer emphasizes the importance of long-term thinking in order to support Canada's research libraries as open access hubs of orderly, sustainable, and productive information.

*_____ 2013. "Libraries, Scholars, and Publishers in Digital Journal and Monograph Publishing." *Scholarly and Research Communication* 4 (1): n.p. http://src-online.ca/index.php/src/article/view/43/118.

Lorimer briefly details the last forty years of scholarly publishing to explicate the current state of affairs. He asserts that a reorganization of the academic publishing infrastructure would greatly encourage forthright contributions to knowledge, especially concerning academic journals and monographs. The splitting of the university press from the university (except in name), coupled with funding cuts and consequent entrepreneurial publishing projects, has hampered the possibilities of academic publishing. If all of the actors of digital scholarly communication—libraries, librarians, scholars on editorial boards, technologically-inclined researchers, programmers, digital humanists, and publishing professionals—were brought together in an inclusive collaboration, digital technology could yield significant benefits for the future of scholarship and knowledge creation.

Losh, Elizabeth. 2012. "Hacktivism and the Humanities: Programming Protest in the Era of the Digital University." In *Debates in the Digital Humanities*, edited by Matthew K. Gold, 161–86. Minneapolis: University of Minnesota Press. http://dhdebates.gc.cuny.edu/debates/text/32.

Losh scans the instantiations of, and relations between, hacktivism and the humanities. She contends, along with scholar Alan Liu, that through an increased self-awareness the digital humanities can actually effect real political, social, public, and institutional change. Losh examines the hacking rhetoric and actions of scholar Cathy Davidson, via the HASTAC collaboratory; the Radical Software Group and its director Alexander Galloway; and the Critical Art Ensemble, with a focus on CAE member and professor Ricardo Dominguez. Losh concludes by acknowledging criticism of the digital humanities, and suggests a solution: digital humanists should engage in more public, political collaborations and conversations.

*Manovich, Lev. 2012. "Trending: The Promises and the Challenges of Big Social Data." In Debates in the Digital Humanities, edited by Matthew K. Gold, 460–75. Minneapolis: University of Minnesota Press. http://dhdebates.gc.cuny.edu/debates/text/15.

Manovich elaborates on the possibilities and limitations of performing humanities research with Big Data. He asserts that although Big Data can be incredibly instructive and useful for humanities work, certain significant roadblocks impede this project. These roadblocks include the fact that only social media companies have access to relevant Big Data; user-generated content is not necessarily authentic, objective, or representative; certain analysis of Big Data requires a level of computer science expertise that humanities researchers do not typically possess; and Big Data is not synonymous with "deep data," the type of data procured through intense, long-term study of subjects. Nevertheless, Manovich looks forward to a future where humanists can overcome these boundaries and integrate Big Data with their research aspirations and projects.

Maxwell, John W. 2014. "Publishing Education in the 21st Century and the Role of the University." *Journal of Electronic Publishing* 17 (2): n.p. http://quod.lib.umich.edu/j/jep/3336451.0017.205?view=text; rgn=main.

From his perspective in the Canadian Institute for Studies in Publishing at Simon Fraser University, Maxwell ruminates on the current state of university-level training in publishing studies, as well as its future role. He considers the shifting economy, and the rise of digital media and practices, as major factors in the current Canadian academic and non-academic publishing scene. Maxwell suggests that the university has a pivotal role to play in reinvigorating publishing by encouraging a supportive community of practice as well as openness to creativity, innovation, and flexibility. Overall, Maxwell underlines the importance of academic publishing studies in the evolving publishing scene.

_____ 2015. "Beyond Open Access to Open Publication and Open Scholarship." *Scholarly and Research Communication* 6 (3): n.p. http:// src-online.ca/index.php/src/article/view/202.

Maxwell considers and comments on the state of scholarly communication. He suggests that online scholarly communication platforms have simply been a means to expedite and render more efficient traditional writing and publishing practices. Maxwell contends that digital scholarly communication, in its current manifestation, is in fact rather conservative. Rather than settling for a traditional and limited production system, academics should move toward more agile, social, and flexible publication modes that consider reader attention and relevance. Maxwell asks scholars and practitioners to reconceive of publishing and publication in the digital age as an opportunity for truly open scholarship.

*McCarty, Willard. 2005. *Humanities Computing*. New York: Palgrave Macmillan.

McCarty examines the field of humanities computing and explores both its limitations and potential. He frames much of his exploration through the mantra that digital humanities can be much more than merely "convenient vending machines for knowledge" (6); the focus must be shifted from automation and delivery to the possibilities for new knowledge creation through digital humanities practices. To this end, McCarty celebrates the tendency toward modelling and manipulation. Drawing heavily on Clifford Geertz's model of/model for theory (and privileging the "model for" concept), McCarty explores how models and unfinished prototypes can be productive spaces of work, knowledge, and play. Models provide invaluable information when they dysfunction, either through inexplicable successes or failures. Of note, he incorporates Martin Heidegger's concept of manipulating the world through technology.

McGregor, Heidi, and Kevin Guthrie. 2015. "Delivering Impact of Scholarly Information: Is Access Enough?" *Journal of Electronic Publishing* 18 (3): n.p. http://quod.lib.umich.edu/j/jep/3336451.0018.302?view =text;rgn=main.

McGregor and Guthrie write on open access from their perspective at ITHA-KA, a not-for-profit organization that focuses on the wide dissemination of knowledge and is most well known for JSTOR, a large-scale digital library service. The authors argue that free access alone is not sufficient for ensuring the broad dissemination, uptake, and impact of knowledge. They then shift focus from access to what they term "productive use," and outline a series of conditions that they deem necessary for a scholarly resource to be considered effective from a knowledge-building perspective. These conditions include literacy, technology, awareness, access, know-how, and training. McGregor and Guthrie conclude that a sustained commitment to these conditions will inevitably heighten scholarly impact and bring the world one step closer to the goal of universal access to knowledge.

Meadows, Alice. 2015. "Beyond Open: Expanding Access to Scholarly Content." *Journal of Electronic Publishing* 18 (3): n.p. http://quod.lib. umich.edu/j/jep/3336451.0018.301?view=text;rgn=main.

Meadows examines the impact of current major public or low-cost open access initiatives worldwide, focusing on the Journal Donation Project,

Research for Life, the International Network for Access to Scientific Papers, patientACCESS, Access to Research, and the Emergency Access Initiative. She emphasizes the wide-ranging value of these initiatives to researchers, professionals, and the public, especially in developing nations or in countries (such as those in the former USSR) that have a history of information bans. Meadows concludes that public or low-cost open access initiatives benefit many different groups of people, including publishers, and should be supported moving forward.

O'Donnell, Daniel, Heather Hobma, Sandra Cowan, Gillian Ayers, Jessica Bay, Marinus Swanepoel, Wendy Merkley, Kelaine Devine, Emma Dering, Inge Genee. 2015. "Aligning Open Access Publication with the Research and Teaching Missions of the Public University: The Case of the Lethbridge Journal Incubator (If 'if's and 'and's were pots and pans)." *Journal of Electronic Publishing* 18 (3): n.p. doi:10.3998/3336451.0018.309. http://quod.lib.umich.edu/j/jep/33 36451.0018.309?view=text;rgn=main.

O'Donnell et al. introduce the Lethbridge Journal Incubator: a project based out of the University of Lethbridge that considers the sustainability of open access publishing in the university. Committing to open access has been a historically difficult choice for academic institutions, as its business model often relies on the complete financial support of the institutions themselves (compared to a subscription system, in which resources can be negotiated across publishers and publications). The Lethbridge Journal Incubator acknowledges the "hidden value" of open access publication, and asks institutions to subsidize this activity in the name of public good as well as to improve the institution's own research and teaching abilities. This model relies on reconceptualizing the scholarly communication production process as one with intrinsic value: for the Lethbridge Journal Incubator, the value lies in the research and training opportunities available for graduate students who are engaged in the project.

Pearce, Nick, Martin Weller, Eileen Scanlon, and Sam Kinsley. 2010. "Digital Scholarship Considered: How New Technologies Could Transform Academic Work." *Education* 16 (1): n.p. http:// ineducation.couros.ca/index.php/ineducation/article/view/44.

Pearce, Weller, Scanlon, and Kinsley establish a definition of digital scholarship: embracing open values and new technologies in order to benefit the academy and society. They explore how digital technologies create new possibilities for open practice. The authors suggest that changes happening in several industries will appear in higher education, either for economic reasons or to adapt to the net-generation of students. They rely on Ernest L. Boyer's dimensions of scholarship from *Scholarship Reconsidered: Priorities of the Professoriate* (1990)—discovery, integration, application, and teaching—in order to discuss the impact of digital technologies. The authors conclude that digital tools can lead to new, open ways of doing scholarship, and call for more research toward establishing effectiveness of the digital tools across disciplines.

Powell, Daniel, Raymond G. Siemens, and William R. Bowen, with Matthew Hiebert and Lindsey Seatter. 2015. "Transformation through Integration: The Renaissance Knowledge Network (ReKN) and a Next Wave of Scholarly Publication." *Scholarly and Research Communication* 6 (2): n.p. http://src-online.ca/index.php/src/article/view/199.

Powell, Siemens, and Bowen, with Hiebert and Seatter, explore the first six months of the Andrew W. Mellon-funded Renaissance Knowledge Network (ReKN). The authors focus on the potential for interoperability and metadata aggregation of various Renaissance and early modern digital projects. They examine how interconnected resources and scholarly environments might integrate publication and markup tools. Powell et al. consider how projects such as ReKN contribute to the shifting practices of contemporary scholarly publishing. For a detailed exploration of the planning phase of ReKN, please see the entry for Powell, Siemens, and the INKE Research Group (2014), immediately below.

Powell, Daniel, Raymond G. Siemens, and the INKE Research Group. 2014. "Building Alternative Scholarly Publishing Capacity: The Renaissance Knowledge Network (ReKN) as Digital Production Hub." *Scholarly and Research Communication* 5 (4): n.p. http://src-online.ca/index.php/src/article/view/183.

Powell, Siemens, et al. report on the status of the Renaissance Knowledge Network (ReKN), an Advanced Research Consortium node. ReKN is a large-scale collaborative project that spans the University of Victoria, the University of Toronto, and Texas A&M University. The authors detail the planning phase of ReKN, a project that aims to centralize and integrate research and production in a single online platform that will serve the specific needs of early modern scholars. Powell and Siemens aim to develop and implement ReKN as a dynamic, holistic scholarly environment. For a further update, please see the entry for Powell, Siemens, and Bowen, with Hiebert and Seatter (2015), above, an article that reflects on the first six months of ReKN development.

*Rosenzweig, Roy. 2006. "Can History Be Open Source? Wikipedia and the Future of the Past." *Journal of American History* **93 (1): 117–46.**
Rosenzweig envisions a model for history scholarship based on the open access, multi-author Wikipedia framework. He concedes that Wikipedia represents an exciting—and perhaps even more ethical—structure of sharing and creating knowledge. Although Rosenzweig thoroughly and comprehensively acknowledges all of the criticisms of Wikipedia from an academic standpoint, he nonetheless proposes that history scholars become more open to incorporating Wikipedia in their scholarly practice. Rosenzweig heralds the many benefits of wiki-based learning and projects for both research and teaching purposes.

San Martin, Patricia Silvana, Paola Caroline Bongiovani, Ana Casali, and Claudia Deco. 2015. "Study on Perspectives Regarding Deposit on Open Access Repositories in the Context of Public Universities in the Central-Eastern Region of Argentina." *Scholarly and Research Communication* **6 (1): n.p. http://src-online.ca/index.php/src/article/view/145.**
San Martin, Bongiovani, Casali, and Deco report on a survey and qualitative study they conducted regarding the dissemination of open access publishing scholarly work. The authors focus on issues related to usability, navigation, and accessibility of institutional repositories in Argentina. For open access institutional repositories, Martin et al. identify a need to manage different types of heterogeneous resources alongside distribution, description, and treatment activities. They consider the barriers to participation in institutional repositories, but suggest that these barriers may be overcome, and that the "physical-virtual campus" (11) that institutional repositories promote and form a part of is possible at a national scale. Overall, the authors see the movement toward broadly instituted open access institutional repositories as a positive and progressive change in the organizational culture of scholarly production in Argentina.

*Siemens, Raymond G. 2002. "Scholarly Publishing at its Source, and at Present." In *The Credibility of Electronic Publishing: A Report to the Humanities and Social Sciences Federation of Canada,* **compiled by Raymond G. Siemens, Michael Best, Elizabeth Grove-White, Alan Burk, James Kerr, Andy Pope, Jean-Claude Guédon, Geoffrey Rockwell, and Lynne Siemens.** *TEXT Technology* **11 (1): 1–128.**
Siemens's introduction to this report focuses on the rethinking of scholarly communication practices in light of new digital forms. He meditates on this topic through the framework of *ad fontes*—the act, or conception, of going

to the source. As he argues, scholars should look at the source or genesis of scholarly communication. For Siemens, the source goes beyond the seventeenth-century inception of the academic print journal to include less formal ways of communicating and disseminating knowledge—i.e., verbal exchanges, epistolary correspondence, and manuscript circulation. In this way, scholars can look past the popular, standard academic journal and into a future of scholarly communication that productively involves varied scholarly traditions and social knowledge practices.

*Vaidhyanathan, Siva. 2002. "The Content-Provider Paradox: Universities in the Information Ecosystem." *Academe* 88 (5): 34–37. doi:10.2307/40252219.
Vaidhyanathan warns against the increasing corporatization of American universities and other knowledge institutions. He argues that universities have begun to commodify knowledge, and that this tactic will eventually lead to the dissolution of the university as a credible source of education. Unfortunately, Vaidhyanathan does not offer an alternative model through which universities can address widespread funding and budget cuts. Nevertheless, taking a similar approach to that of Willard McCarty in *Humanities Computing*, Vaidhyanathan reminds his readers that education is not simply information, and should not be treated (or sold) as such.

Vandendorpe, Christian. 2015. "Wikipedia and the Ecosystem of Knowledge." *Scholarly and Research Communication* 6 (3): n.p. http://src-online.ca/index.php/src/article/view/201.
Vandendorpe argues for the broad uptake of Wikipedia across the academy, contending that researchers need to edit on Wikipedia and to share their specialized knowledge with the rest of the world. In this way, Vandendorpe argues, scholars can easily share their findings broadly and publicly. He emphasizes online, popular, and open access environments in the growing media ecology that supports scholarly communication. Vandendorpe champions the opportunities afforded by serious academic engagement with Wikipedia.

Van de Sompel, Herbert, Sandy Payette, John Erickson, Carl Lagoze, and Simeon Warner. 2004. "Rethinking Scholarly Communication: Building the System that Scholars Deserve." *D-Lib Magazine* 10 (9): n.p. doi:10.1045/september2004-vandesompel. http://www.dlib.org/dlib/september04/vandesompel/09vandesompel.html.
Van de Sompel, Payette, Erickson, Lagoze, and Warner ruminate on transforming scholarly communication to better serve and facilitate knowledge creation.

They primarily target the current academic journal system, which they see as constraining scholarly work because it is expensive, difficult to access, and print-biased. The authors propose a digital system for scholarly communication that more accurately incorporates ideals of interoperability, adaptability, innovation, documentation, and democratization. This proposed system would be implemented as a concurrent knowledge production environment instead of as a mere stage, annex, or afterthought for scholarly work.

Van House, Nancy A. 2003. "Digital Libraries and Collaborative Knowledge Construction." In *Digital Library Use: Social Practice in Design and Evaluation*, edited by Ann Peterson Bishop, Nancy A. Van House, and Barbara P. Buttenfield, 271–95. Cambridge, MA: MIT Press.

Van House reminds her readers that libraries are more than just storehouses; libraries comprehensively support and foster knowledge creation. Consequently, Van House claims, designing and building effective digital libraries depends on a thorough understanding of knowledge work. For Van House, the emergence of digital libraries represents a significant shift in how individuals and communities create knowledge. Digital libraries often foster transgressive, situated, distributed, and social networks of research and knowledge production. Notably, she reinforces the concept that artifacts are not knowledge in and of themselves; knowledge is a complex social phenomenon rooted in contact, daily practice, and partial mediation by artifacts. As such, digital libraries function differently than as mere conduits— digital libraries are boundary objects, and they affect knowledge work significantly by introducing variation in terms of manipulability, credibility, inscription, access, and organization.

10. Social Knowledge Creation in Electronic Journals and Monographs

*Biagioli, Mario. 2002. "From Book Censorship to Academic Peer Review." *Emergences* 12 (1): 11–45. doi:10.1080/1045722022000003435.

Biagioli details the historical and epistemological shifts that have led to the academic peer review system as it is now known. Contrary to its contemporary role, peer review began as an early modern disciplinary technique closely related to book censorship and required for social and scholarly certification of institutions and individuals alike. The rise of academic journals shifted this constrained and royally-mandated position; no longer a self-sustaining system of judgment and reputation dictated by a small group of identified and accredited professionals, (often blind) peer review now focuses on disseminating knowledge and scholarship to the wider community. Biagioli also states that journals have moved from officially representing specific academic institutions to being community owned and operated, as responsibilities, duties, and readership are now dispersed among a group of like-minded scholars.

*Borgman, Christine L. 2007. *Scholarship in the Digital Age: Information, Infrastructure, and the Internet.* **Cambridge, MA: MIT Press.**
Borgman lays out research questions and hypotheses concerning the evolving scholarly infrastructure and modes of communication in the digital environment. She deduces that the inherent social elements of scholarship endure, despite new technologies that alter significantly the way scholarship is performed, disseminated, and archived. Scholarship and scholarly activities continue to exist in a social network of varying actors and priorities. Notably, Borgman focuses on the "data deluge"—the increasing amount of generated data and data accessed for research purposes. The influences of large data sets, as well as how these data sets will be preserved in keeping with library and archival conventions, are subjects of particular significance. Overall, Borgman synthesizes the various aspects of contemporary scholarship, and reflects on the increasingly pervasive digital environment.

Christie, Alex, and the INKE and MVP Research Groups. 2014. "Interdisciplinary, Interactive, and Online: Building Open Communication Through Multimodal Scholarly Articles and Monographs." *Scholarly and Research Communication* **5 (4): n.p. http://src-online.ca/index. php/src/article/view/190.**
Christie considers the possibilities for uniting text-based scholarship with multimodal content. He focuses on features and platforms that are suitable for both text-based and multimedia scholarship, and suggests that digital scholarly publishing may better facilitate interaction between humanities scholars and the public. For Christie, rethinking scholarly communication in these ways must be supported by advanced cyberinfrastructure. The knowledge products and environments that result must also privilege multimedia, interactivity, user engagement, and implementation. This sort of platform thinking inheres a strategic reconsideration of interactivity, interdisciplinarity, design, and infrastructure investment.

Cohen, Daniel J. 2012. "The Social Contract of Scholarly Publishing." In *Debates in the Digital Humanities,* **edited by Matthew K. Gold, 319–21. Minneapolis: University of Minnesota Press. http://dhdebates. gc.cuny.edu/debates/text/27.**
Cohen remarks on the social contract of scholarly publishing—the contract between the producers (authors, editors, publishers) and the consumers

(readers), or the "supply side" and the "demand side." According to Cohen, individuals on the supply side have become increasingly experimental in recent years, but there has not been enough attention paid to the demand side. Cohen asserts that a thorough consideration of the demand side is necessary for the social contract to endure into the digital age. To accomplish this, academics must think more socially and become increasingly cognizant of the design, packaging, and outreach of their publishing ventures.

*Fitzpatrick, Kathleen. 2012. "Beyond Metrics: Community Authorization and Open Peer Review." In *Debates in the Digital Humanities,* edited by Matthew K. Gold, 452–59. Minneapolis: University of Minnesota Press. http://dhdebates.gc.cuny.edu/debates/text/7.

Fitzpatrick calls for a reform of scholarly communication via open peer review. She argues that the Internet has provoked a conceptual shift wherein (textual) authority is no longer measured by a respected publisher's stamp; rather, she contends, authority is now located in the community. As concepts of authority change and evolve in the digital sphere, so should methods. Peer review should be opened to various scholars in a field, as well as to non-experts from other fields and citizen scholars. Fitzpatrick claims that this sort of crowdsourcing of peer review could more accurately represent scholarly and non-scholarly reaction, contribution, and understanding. Digital humanities and new media scholars already have the tools to measure digital engagement with a work; now, a better model of peer review should be implemented to take advantage of the myriad, social, networked ways scholarship is (or could be) produced.

*_____ 2009. "Peer-To-Peer Review and the Future of Scholarly Authority." *Cinema Journal* 48 (2): 124–29. doi:10.1353/cj.0.0095.

Fitzpatrick explains that decentralized and displaced authority structures are taking over scholarly communication, and intellectual authority is shifting to spaces such as Wikipedia. Scholars must therefore embrace similarly open structures and public access, or the academic world will appear divorced from real world practices. For this reason, online peer-reviewed journals should not follow print practices of peer review, but must adapt and shape a new scholarly system. Current peer review processes do not necessarily ensure that the best work is in circulation, and in fact tend to replicate hierarchical and status-based privileges. Fitzpatrick argues for open process, web native modes of peer review in a peer-to-peer structure. Finally, she advocates for the need to articulate these values and standards to credentialing bodies in order for a more appropriate model of intellectual authorization to emerge.

*_____ 2011. *Planned Obsolescence: Publishing, Technology, and the Future of the Academy.* New York: New York University Press. ("Introduction: Obsolescence," 1–14, and Chapter 3: "Texts," 89–120, are accessible at http://raley.english.ucsb.edu/wp-content2/uploads/234/Fitzpatrick.pdf.)

Fitzpatrick duly surveys and calls for a reform of academic publishing. She argues for more interactivity, communication, and peer-to-peer review, as well as a significant move toward digital scholarly publishing. Fitzpatrick demonstrates that the current mode of scholarly publishing is economically unviable. Moreover, tenure and promotion practices based primarily on traditional modes of scholarly publishing need to be reformed. Fitzpatrick acknowledges certain touchstones of the academy (peer review, scholarship, sharing ideas), and how these tenets have been overshadowed by priorities shaped, in part, by mainstream academic publishing practices and concepts. She details her own work with CommentPress, and the benefits of publishing online with an infrastructure that enables widespread dissemination as well as concurrent reader participation via open peer review.

*Fjällbrant, Nancy. 1997. "Scholarly Communication—Historical Development and New Possibilities." In *Proceedings of the IATUL Conferences.* West Lafayette, IN: Purdue University Libraries e-Pubs. http://docs.lib.purdue.edu/cgi/viewcontent.cgi?article=1389&context=iatul.

In order to study the widespread transition of scholarly communication from print to electronic formats, Fjällbrant details the history of the scientific journal. Academic journals had emerged in seventeenth-century Europe, and the first of these, the *Journal des Sçavans,* was published in 1665 in Paris. The first learned societies formed at this time—the Royal Society in London and the Académie des Sciences in Paris—were primarily concerned with the dissemination of knowledge, and the scholarly journal developed out of a desire by researchers to share their findings with others in a cooperative forum. Following the lead of the Royal Society, some of whose members had read the *Journal des Sçavans,* other societies established similar serial publications. Although there were other contemporaneous forms of scholarly communication, including the letter, the scientific book, the newspaper, and the cryptic anagram system, the journal emerged as a primary source of scholarly communication. It met the needs of various stakeholders: the general public, booksellers and publishers, libraries, authors who wished to make their work public and claim ownership, the scientific community invested in reading and applying other scientists' findings, and academic institutions that required metrics for evaluating faculty.

*Guédon, Jean-Claude. 2008. "Digitizing and the Meaning of Knowledge." *Academic Matters* (October–November): 23–26. http://www.academicmatters.ca/assets/AM_SEPT'08.pdf.

Guédon briefly sketches the recent history of scholarly communication and publishing and meditates on alternatives to the current state of affairs. He concludes that although open source publishing is a relatively recent phenomenon, it adroitly embodies the ethos and traditional practices of scholarship (especially in the sciences). For Guédon, open source publishing represents the open, endless appropriation of knowledge and discipline-wide conversation that has traditionally defined academic work. Guédon champions this move toward open, shared knowledge versus the continued exploitation of academics, librarians, and universities by the large corporate publishing companies currently relied upon for scholarly communication and accreditation.

Guldi, Jo. 2013. "Reinventing the Academic Journal." In *Hacking the Academy: New Approaches to Scholarship and Teaching from Digital Humanities*, edited by Daniel J. Cohen and Tom Scheinfeldt, 19–24. Ann Arbor: University of Michigan Press. doi:10.3998/dh.12172434.0001.001.

Guldi calls for a rethinking of scholarly journal practices in light of the emergence and allowances of Web 2.0. She argues that journals can re-establish themselves as forthright facilitators of knowledge creation if they adopt notions of interoperability, curation, multimodal scholarship, open access, networked expertise, and transparency regarding review and timelines. For Guldi, the success of the academic journal depends on incorporating social bookmarking tools and wiki formats. Journals should assume a progressive attitude predicated on sharing and advancing knowledge instead of a limiting view based on exclusivity, profit, and intellectual authority.

Liu, Alan. 2009. "The End of the End of the Book: Dead Books, Lively Margins, and Social Computing." *Michigan Quarterly Review* 48 (4): 499–520.

Liu argues that books have always, in a sense, been social media. He acknowledges the increase in bibliographic and material textual studies and the correspondences between new digital reading environments and the book, with a focus on paratextual materials and marginality. In this way, Liu contests apocalyptic claims of the death of the book. Notably, Liu channels his assertions through an analysis of humanities-based digital research projects: Collex, Open Journal Systems, and PreE. He suggests that these environments allow for more thoughtful online engagement and user operability (the capacity to effectively and easily manipulate and tailor research practices) than their mainstream counterparts. The trend toward reading, researching, and writing in digital spaces does not herald the end of the book; rather, certain digital humanities projects are synthesizing integral reading practices in order to improve and facilitate more widespread knowledge production, with an eye to the inherent sociality of texts.

*Lorimer, Rowland. 2013. "Libraries, Scholars, and Publishers in Digital Journal and Monograph Publishing." *Scholarly and Research Communication* 4 (1): n.p. http://src-online.ca/index.php/src/article/view-File/43/117.

Lorimer briefly details the last forty years of scholarly publishing to explicate the current state of affairs. He asserts that a reorganization of the academic publishing infrastructure would greatly encourage forthright contributions to knowledge, especially concerning academic journals and monographs. The splitting of the university press from the university (except in name), coupled with funding cuts and consequent entrepreneurial publishing projects, has hampered the possibilities of academic publishing. If all of the actors of digital scholarly communication—libraries, librarians, scholars on editorial boards, technologically-inclined researchers, programmers, digital humanists, and publishing professionals—were brought together in an inclusive collaboration, digital technology could yield significant benefits for the future of scholarship and knowledge creation.

*O'Donnell, Daniel, Heather Hobma, Sandra Cowan, Gillian Ayers, Jessica Bay, Marinus Swanepoel, Wendy Merkley, Kelaine Devine, Emma Dering, and Inge Genee. 2015. "Aligning Open Access Publication with the Research and Teaching Missions of the Public University: The Case of the Lethbridge Journal Incubator (If 'if's and 'and's were pots and pans)." *Journal of Electronic Publishing* 18 (3): n.p. doi:10.3998/3336451.0018.309. http://quod.lib.umich.edu/j/jep/33 36451.0018.309?view=text;rgn=main.

O'Donnell et al. introduce the Lethbridge Journal Incubator: a project based out of the University of Lethbridge that considers the sustainability of open access publishing in the university. Committing to open access has been a historically difficult choice for academic institutions, as its business model often relies on the complete financial support of the institutions themselves (compared to a subscription system, in which resources can be negotiated across publishers and publications). The Lethbridge Journal Incubator acknowledges the "hidden value" of open access publication, and asks institutions to subsidize this activity in the name of public good as well as to improve the institution's own research and teaching abilities. This model relies on re-conceptualizing the scholarly communication production process as one with intrinsic value: for the Lethbridge Journal Incubator, the value lies in the research and training opportunities available for graduate students who are engaged in the project.

*Siemens, Raymond G. 2002. "Scholarly Publishing at its Source, and at Present." In The *Credibility of Electronic Publishing: A Report to the Humanities and Social Sciences Federation of Canada*, compiled by Raymond G. Siemens, Michael Best, Elizabeth Grove-White, Alan Burk, James Kerr, Andy Pope, Jean-Claude Guédon, Geoffrey Rockwell, and Lynne Siemens. TEXT *Technology* 11 (1): 1–128.

Siemens's introduction to this report focuses on the rethinking of scholarly communication practices in light of new digital forms. He meditates on this topic through the framework of *ad fontes*—the act, or conception, of going to the source. As he argues, scholars should look at the source or genesis of scholarly communication. For Siemens, the source goes beyond the seventeenth-century inception of the academic print journal to include less formal ways of communicating and disseminating knowledge—i.e., verbal exchanges, epistolary correspondence, and manuscript circulation. In this way, scholars can look past the popular, standard academic journal and into a future of scholarly communication that productively involves varied scholarly traditions and social knowledge practices.

*Van de Sompel, Herbert, Sandy Payette, John Erickson, Carl Lagoze, and Simeon Warner. 2004. "Rethinking Scholarly Communication: Building the System that Scholars Deserve." *D-Lib Magazine* 10 (9): n.p. doi:10.1045/september2004-vandesompel. http://www.dlib. org/dlib/september04/vandesompel/09vandesompel.html.

Van de Sompel, Payette, Erickson, Lagoze, and Warner ruminate on transforming scholarly communication to better serve and facilitate knowledge creation. They primarily target the current academic journal system, which they see as constraining scholarly work because it is expensive, difficult to access, and print-biased. The authors propose a digital system for scholarly communication that more accurately incorporates ideals of interoperability, adaptability, innovation, documentation, and democratization. This proposed system would be implemented as a concurrent knowledge production environment instead of as a mere stage, annex, or afterthought for scholarly work.

11. Social Knowledge Creation in Electronic Scholarly Editions and e-Books

Aarseth, Espen J. 1997. "Introduction." In *Cybertext: Perspectives on Ergodic Literature*, 1–23. Baltimore: Johns Hopkins University Press.
Aarseth attempts to develop a theory of cybertext works, with a focus on "ergodic texts." Aarseth's scholarly interest lies in texts that are purposefully shaped by the reader's tangible and visible actions and decisions. He bases his speculation on the concept that cybertexts are labyrinthine and user dependent, and contain feedback loops. Aarseth criticizes the counterarguments that many texts can be read as cybertexts, but does not concede that this distinction derives from cybertexts' necessarily electronic mode. The inherent performativity involved in reading cybertexts occurs in a network of various parts and participants, compared with the more conventional reading model of reader/author/text. Further, Aarseth argues, ergodic texts (primarily virtual games and multi-user domains [MUDs]) are defined by the agency and authority of the human subject (reader) whose decisions affect the outcome of the text as a whole.

Andersen, Christian Ulrik, and Søren Bro Pold. 2014. "Post-digital Books and Disruptive Literary Machines." *Formules/Revue des Créations Formelles et Littératures à Contraintes* 18: 169–88.
Andersen and Pold conclude that the book is now "post-digital," and they provide various examples of innovative and common textual artifacts to support this claim. They argue that the infrastructure around electronic publications has been normalized and integrated fully into international reading, writing, and consumption practices. Andersen and Pold emphasize the capitalism inherent in current mainstream digital text platforms, such as Amazon. The authors detail and vouch for attempts to counter the controlled, corporate, and user-objectifying electronic text ecosystem.

Bath, Jon, and Scott Schofield. 2015. "The Digital Book." In *The Cambridge Companion to the History of the Book*, edited by Leslie Howsam, 181–95. Cambridge: Cambridge University Press.
Bath and Schofield reflect on the rise of the e-book by contemplating the various moving parts involved in its history and production. They focus on, and contribute to, the scholarly engagement with e-books, and they provide a comprehensive survey of theorists, including Johanna Drucker, Elizabeth Eisenstein, N. Katherine Hayles, Matthew Kirschenbaum, Jerome McGann, D.F. McKenzie, and Marshall McLuhan. Bath and Schofield integrate these theorists into a larger argument that suggests that both a nuanced understanding of book history and a comprehensive familiarity with digital

scholarship is necessary to fully grasp the material and historical significance of the e-book. The authors conclude with a call to book history and digital humanities specialists (a.k.a. "scholar-coders") to collaborate and develop new digital research environments together.

*Christie, Alex, and the INKE and MVP Research Groups. 2014. "Interdisciplinary, Interactive, and Online: Building Open Communication Through Multimodal Scholarly Articles and Monographs." *Scholarly and Research Communication* 5 (4): n.p. http://src-online.ca/index.php/src/article/view/190.

Christie considers the possibilities for uniting text-based scholarship with multimodal content. He focuses on features and platforms that are suitable for both text-based and multimedia scholarship, and suggests that digital scholarly publishing may better facilitate interaction between humanities scholars and the public. For Christie, rethinking scholarly communication in these ways must be supported by advanced cyberinfrastructure. The knowledge products and environments that result must also privilege multimedia, interactivity, user engagement, and implementation. This sort of platform thinking inheres a strategic reconsideration of interactivity, interdisciplinarity, design, and infrastructure investment.

*Clement, Tanya. 2011. "Knowledge Representation and Digital Scholarly Editions in Theory and Practice." *Journal of the Text Encoding Initiative* 1: n.p. doi:10.4000/jtei.203.

Clement reflects on scholarly digital editions as sites of textual performance, wherein the editor lays down and privileges various narrative threads for the reader to pick up and interpret. She underscores this theoretical discussion with examples from her own work with the digital edition *In Transition: Selected Poems by the Baroness Elsa von Freytag-Loringhoven*, as well as TEI and XML encoding and the Versioning Machine. Clement details how editorial decisions shape the social experience of an edition. By applying John Bryant's theory of the fluid text to her own editorial practice, she focuses on concepts of various textual performances and meaning-making events. Notably, Clement also explores the idea of the social text network. She concludes that the concept of the network is not new to digital editions; nevertheless, conceiving of a digital edition as a network of various players, temporal spaces, and instantiations promotes fruitful scholarly exploration.

*Cohen, Daniel J. 2012. "The Social Contract of Scholarly Publishing." In *Debates in the Digital Humanities*, edited by Matthew K. Gold, 319–21.

Minneapolis: University of Minnesota Press. http://dhdebates.gc.cuny.edu/debates/text/27.

Cohen remarks on the social contract of scholarly publishing—the contract between the producers (authors, editors, publishers) and the consumers (readers), or the "supply side" and the "demand side." According to Cohen, individuals on the supply side have become increasingly experimental in recent years, but there has not been enough attention paid to the demand side. Cohen asserts that a thorough consideration of the demand side is necessary for the social contract to endure into the digital age. To accomplish this, academics must think more socially and become increasingly cognizant of the design, packaging, and outreach of their publishing ventures.

Crompton, Constance, Raymond G. Siemens, and Alyssa Arbuckle, with the INKE Research Group. 2015. "Enlisting 'Vertues Noble & Excelent': Behavior, Credit, and Knowledge Organization in the Social Edition." *Digital Humanities Quarterly* 9 (2): n.p. http://www.digitalhumanities.org/dhq/vol/9/2/000202/000202.html.

Crompton, Siemens, and Arbuckle consider the gender factors involved in social editions by drawing on their experience developing a Wikibook edition of the *Devonshire Manuscript*, a sixteenth-century multi-author verse miscellany. The authors argue that while the Wikimedia suite can often devolve into openly hostile online spaces, Wikimedia projects remain important for the contemporary circulation of knowledge. The key, for the authors, is to encourage gender equity in social behaviour, credit sharing, and knowledge organization in Wikimedia, rather than abandoning it in favour of a more controlled collaborative environment for edition production and dissemination.

Fitzpatrick, Kathleen. 2007. "CommentPress: New (Social) Structures for New (Networked) Texts." *Journal of Electronic Publishing* 10 (3): n.p. doi:10.3998/3336451.0010.305.

Fitzpatrick meditates on the current state and future possibilities of electronic scholarly publishing. She focuses her consideration on a study of Comment-Press, a digital scholarly publishing venue that combines the hosting of long texts with social network features. Fitzpatrick argues that community and collaboration are at the heart of scholarly knowledge creation—or at least, they should be. Platforms such as CommentPress acknowledge the productive capabilities of scholarly collaboration, and promote this fruitful interaction between academics. Although Fitzpatrick admits that CommentPress is not the only or best answer to the questions of shifting scholarly communication, she celebrates its emergence as a service for the social interconnection and knowledge production of authors and readers in an academic setting.

*_____ 2011. *Planned Obsolescence: Publishing, Technology, and the Future of the Academy*. New York: New York University Press. ("Introduction: Obsolescence," 1–14, and Chapter 3: "Texts," 89–120, are accessible at http://raley.english.ucsb.edu/wp-content2/uploads/234/Fitzpatrick.pdf.)

Fitzpatrick duly surveys and calls for a reform of academic publishing. She argues for more interactivity, communication, and peer-to-peer review, as well as a significant move toward digital scholarly publishing. Fitzpatrick demonstrates that the current mode of scholarly publishing is economically unviable. Moreover, tenure and promotion practices based primarily on traditional modes of scholarly publishing need to be reformed. Fitzpatrick acknowledges certain touchstones of the academy (peer review, scholarship, sharing ideas), and how these tenets have been overshadowed by priorities shaped, in part, by mainstream academic publishing practices and concepts. She details her own work with CommentPress, and the benefits of publishing online with an infrastructure that enables widespread dissemination as well as concurrent reader participation via open peer review.

Flanders, Julia. 2005. "Detailism, Digital Texts, and the Problem of Pedantry." *TEXT Technology* 14 (2): 41–70. http://texttechnology.mcmaster.ca/pdf/vol14_2/flanders14-2.pdf.

Flanders acknowledges the long-standing academic anxiety surrounding detailism, automation, and numerical or scientific applications in textual studies and literary criticism. She contends that text analysis and digital editing should not be written off as reductionist or unimportant; rather, these humanities computing practices open up new fields of play and reader-based engagement and interpretation. Flanders argues that text analysis practitioners and scholarly digital editors are very aware of the consequences and nature of their work. Contrary to critics' perspectives, these scholars do not consider computation the be-all and end-all of scholarship: computation is a means of expediting minute and tedious tasks in order to further—and differentiate—interpretation and knowledge creation.

Jankowski, Nicholas W., Andrea Scharnhorst, Clifford Tatum, and Zuotian Tatum. 2013. "Enhancing Scholarly Publications: Developing Hybrid Monographs in the Humanities and Social Sciences." *Scholarly and Research Communication* 4 (1): n.p. http://src-online.ca/index.php/src/article/viewFile/40/123.

Jankowski, Scharnhorst, Tatum, and Tatum present a report on their project, "Enhancing Scholarly Publishing in the Humanities and Social Sciences" (2011), an attempt to develop interlinked, comprehensive digital versions of humanities and social sciences texts. They detail their guiding activities and principles for developing enhanced scholarly publications, and explain how they enacted these principles: providing identifiers and citation information; using popular file formats; achieving adequate technical quality; considering legal issues; addressing availability and sustainability; considering ownership and responsibility; indicating peer review or ranking; balancing complexity with utility; and demonstrating the connection between objects. For Jankowski et al., this project revealed the need to extend the theoretical understanding of current scholarly publishing transformations. They recommend the development of an empirical research agenda that relates to such a theoretical understanding.

*Liu, Alan. 2009. "The End of the End of the Book: Dead Books, Lively Margins, and Social Computing." *Michigan Quarterly Review* **48 (4): 499–520.**

Liu argues that books have always, in a sense, been social media. He acknowledges the increase in bibliographic and material textual studies and the correspondences between new digital reading environments and the book, with a focus on paratextual materials and marginality. In this way, Liu contests apocalyptic claims of the death of the book. Notably, Liu channels his assertions through an analysis of humanities-based digital research projects: Collex, Open Journal Systems, and PreE. He suggests that these environments allow for more thoughtful online engagement and user operability (the capacity to effectively and easefully manipulate and tailor research practices) than their mainstream counterparts. The trend toward reading, researching, and writing in digital spaces does not herald the end of the book; rather, certain digital humanities projects are synthesizing integral reading practices in order to improve and facilitate more widespread knowledge production, with an eye to the inherent sociality of texts.

McGann, Jerome. 2006. "From Text to Work: Digital Tools and the Emergence of the Social Text." *TEXT: An Interdisciplinary Annual of Textual Studies* **16: 49–62. http://www.jstor.org/stable/30227956.**

McGann meditates on the possibilities that digital editing affords for instantiations of social textuality. He argues that well designed digital editions bring significant opportunities for the social text (as bibliography scholar D.F. McKenzie championed). In contrast to their more conventional predecessors, digital editions can more accurately represent the dynamic relations inherent in the production and reception of a text. By simulating bibliographical and socio- textual phenomena, and employing carefully designed user interfaces that allow for multiple or specialized readings, digital editions can better represent texts as social artifacts and reading as a social act.

_____ 1991. *The Textual Condition.* **Princeton, NJ: Princeton University Press.**

McGann persuasively argues that the meaning of texts derives from the use of texts. As embodied phenomena, texts are always more expansive and inclusive than mere form or mere content. According to McGann, literary texts are social experiences, socially made, and thus require a form of social editing. McGann examines various theories and schools of textual editing, including literary theorist Gerard Genette's conception of the paratextual apparatus. Further, McGann argues that the concept of "authorial intention" is a fallacy; texts pass through various hands—even through the author's hands more than once—and to isolate one original, authentic, or "true" version is a technical and conceptual impossibility.

Moretti, Franco. 2005. *Graphs, Maps, Trees: Abstract Models for a Literary History.* **London and New York: Verso.**

Moretti develops his theory on "distant reading," the practice of interpreting literature by looking at large scale patterns—namely through using graphs, maps, and trees as analytical tools. Moretti criticizes literary studies for having too narrow or too close of a focus on specific, canonical literary works, and thus missing significant themes and trends. He draws on various analyses, from graphs of book production in the eighteenth century to geometric maps/diagrams of village stories to Darwinian theories of diverging evolution. Moretti concludes that distant reading can open up literary studies to a more morphological and inclusive way of analyzing and making knowledge.

Robinson, Peter. 2010. "Electronic Editions for Everyone." In *Text and Genre in Reconstruction: Effects of Digitalization on Ideas, Behaviours, Products and Institutions,* **edited by Willard McCarty, 145–63. Cambridge: Open Book Publishers.**

Robinson acknowledges the significant gap between the projected success of digital editions at their inception and the actual popularity of these editions now. He suggests that editors significantly alter their methods of digital edition creation in order to reflect on and take advantage of increasingly sophisticated technology and Web 2.0 practices. Robinson claims that digital editions would gain popularity if they were modelled in a more fluid and distributed form. He argues that editors should move away from compiling scholarly digital editions in a dedicated space with a specific interface, method of organization, and formally delineated content; instead, they should develop Internet applications that track a user's research interests and practices and automatically compile relevant information. This method would substantially alter digital scholarship and reflect the networked realm of the Internet much more accurately— and perhaps with more ease—than current digital editions are capable of.

*Saklofske, Jon, with the INKE Research Group. 2012. "Fluid Layering: Reimagining Digital Literary Archives Through Dynamic, User-Generated Content." *Scholarly and Research Communication* 3 (4): n.p. http://src-online.ca/index.php/src/article/viewFile/70/181.

Saklofske argues that while the majority of print and digital editions exist as isolated collections of information, changing practices in textual scholarship are moving toward a new model of production. He uses the example of *NewRadial*, a prototype information visualization application, to showcase the potential of a more active public archive. Specifically, Saklofske focuses on making room for user-generated data that transforms the edition from a static repository into a dynamic and co-developed space. He champions the argument that the digital archive should place user-generated content in a more prominent position through re-imagining the archive as a site of critical engagement, dialogue, argument, commentary, and response. In closing, Saklofske poses five open-ended questions to the community at large as a way of kickstarting a conversation regarding the challenges of redesigning the digital archive.

Saklofske, Jon, and Jake Bruce, with the INKE Modelling and Prototyping Team and the INKE Team. 2013. "Beyond Browsing and Reading: The Open Work of Digital Scholarly Editions." *Scholarly and Research Communication* 4 (3): n.p. http://src-online.ca/index.php/src/article/view/119.

Saklofske and Bruce discuss *NewRadial*, an Implementing New Knowledge Environments (INKE) prototype scholarly edition environment. The prototype draws together primary texts, secondary scholarship, and related knowledge communities into a digital scholarly edition. Saklofske and Bruce deem this a social edition. *NewRadial* provides an open, shared workspace where users may explore, sort, group, annotate, and contribute to secondary scholarship creation collaboratively.

Shillingsburg, Peter. 2006. *From Gutenberg to Google: Electronic Representations of Literary Texts*. Cambridge: Cambridge University Press.

Shillingsburg ruminates on editorial practice and his ideal digital edition: the "knowledge site." A knowledge site, in Shillingsburg's conception, is a space where multiple editions of a text could be combined in a straightforward manner. Based on his experience and knowledge of editorial practice and the mandates of the scholarly edition, he deems various elements necessary for a knowledge site, including: basic and inferred data, internal links, bibliographical analysis, contextual data, intertextuality, linguistic analysis, reception history, and adaptations. Furthermore, in keeping with the notion that digital scholarly editions have the capacity to shift the possession of the text to the users, Shillingsburg would ideally include opportunities for user- generated markup, variant texts, explanatory notes and commentary, and a personal note space. Concurrently, Shillingsburg argues that editing is never neutral, but rather an interference in the history and status of the text. The overt acknowledgement of the intrusive nature of editing is imperative for all successful scholarly editions. Since unobtrusive editing and universal texts are non-existent, scholarly editions are better conceived of as select interpretations of texts for specific means.

*Siemens, Raymond G., Meagan Timney, Cara Leitch, Corina Koolen, and Alex Garnett, with the ETCL, INKE, and PKP Research Groups. 2012. "Toward Modeling the *Social* Edition: An Approach to Understanding the Electronic Scholarly Edition in the Context of New and Emerging Social Media." *Digital Scholarship in the Humanities* (formerly *Literary and Linguistic Computing*) 27 (4): 445–61. doi:10.1093/llc/fqs013.

Siemens, Timney, Leitch, Koolen, Garnett, et al. present a vision of an emerging manifestation of the scholarly digital edition: the social edition. The authors ruminate on both the potential and already-realized intersections between scholarly digital editing and social media. For Siemens et al., many scholarly digital editions do not readily employ the collaborative electronic tools available for use in a scholarly context. The authors seek to remediate this lack of engagement, especially concerning opportunities to integrate collaborative annotation, user-derived content, folksonomy tagging, community bibliography, and text analysis capabilities within a digital edition. Furthermore, Siemens et al. envision the conceptual role of the editor—traditionally a single authoritative individual—as a reflection of facilitation rather than of didactic authority. A social edition predicated on these shifts and amendments would allow for increased social knowledge creation by a community of readers and scholars, academics and citizens alike.

Smith, Martha Nell. 2004. "Electronic Scholarly Editing." In *A Companion to Digital Humanities*, edited by Susan Schreibman, Raymond G. Siemens, and John Unsworth, 306–22. Oxford: Blackwell.

Smith relies on her experience with the Dickinson Electronic Archives to formulate a conceptual theory of and argument for electronic scholarly editing. For Smith, a significant benefit of the digital scholarly edition is the shift from unilateral authority to networked experience, from the voice of the sole editor to the polyphonic interpretation of multiple readers. Smith acknowledges the various elements that allow for social knowledge production in the digital scholarly edition, including: comprehensive inclusion of various artifacts and digital surrogates; enabling of multiple editorial theories and consequent readings; engagement of many editorial and readerly intentions and priorities; and social communication via readers' responses, preferences, and tailored readings. Smith concludes that electronic scholarly editing offers the opportunity for more inclusive and democratic knowledge production.

Stein, Bob. 2015. "Back to the Future." *Journal of Electronic Publishing* 18 (2): n.p. doi:10.3998/3336451.0018.204.

Stein considers the digital book as a *place* rather than an object or tool—a place where readers gather, socially. He details the experiments at his Institute for the Future of the Book with social platforms, including creating an online social edition of Mackenzie Wark's *Gamer Theory*, and their current work with SocialBook. SocialBook is an online, collaborative reading platform that encourages readers to comment on the text and interact with each other. Stein makes reference to historic social reading practices, and infers that platforms such as SocialBook are closely aligned to these traditions.

Vandendorpe, Christian. 2012. "Wikisource and the Scholarly Book." *Scholarly and Research Communication* 3 (4): n.p. http://src-online.ca/src/index.php/src/article/viewFile/58/146.

Vandendorpe contemplates Wikisource, a project of the Wikimedia Foundation, as a potential platform for reading and editing scholarly books. He comes to this conclusion after considering what the ideal e-book or digital knowledge environment should look like. For Vandendorpe, this artifact must be available on the web; reflect the metaphor of a forest of knowledge, rather than a container; situate the reader at the centre of the experience; and be open, reliable, robust, and expandable. Wikisource, the author concludes, has the potential to meet these criteria: it enables quality editing and robust versioning, and has various display options. That being said, Vandendorpe also outlines areas of development necessary for Wikisource to become an ideal candidate for this sort of knowledge creation.

*Vetch, Paul. 2010. "From Edition to Experience: Feeling the Way towards User-Focused Interfaces." In *Electronic Publishing: Politics and Pragmatics*, edited by Gabriel Egan, 171–84. New Technologies in Medieval and Renaissance Studies 2. Tempe, AZ: Iter Inc., in collaboration with the Arizona Center for Medieval and Renaissance Studies.

Vetch explores the nuances of a user-focused approach to scholarly digital projects, arguing that the prevalence of Web 2.0 practices and standards requires scholars to rethink the design of scholarly digital editions. For Vetch, editorial teams need to shift their focus to questions concerning the user. For instance, how will users customize their experience of the digital edition? What new forms of knowledge can develop from these interactions? Moreover, how can rethinking the interface design of scholarly digital editions promote more user engagement and interest? Vetch concludes that a user-focused approach is necessary for the success of scholarly publication in a constantly shifting digital world.

12. Exemplary Instances of Social Knowledge Construction

*Cohen, Daniel J. 2008. "Creating Scholarly Tools and Resources for the Digital Ecosystem: Building Connections in the Zotero Project." *First Monday* 13 (8): n.p. doi:10.5210/fm.v13i8.2233.

Cohen details how the Zotero project exemplifies both a Web 2.0 and a traditional scholarly ethos. He conceptualizes Zotero as a node in an interconnected digital ecosystem that builds bridges instead of hoarding information. Zotero is a widely used, open source, community-based bibliography tool. It exists on top of the browser as an extension, has maintained an API since its inception, and boasts comprehensive user features. As an easy-to-use collaborative tool, Zotero acts as both an effective scholarly resource and a facilitator of social knowledge creation.

*Cohen, Daniel J., and Tom Scheinfeldt. 2013. "Preface." In *Hacking the Academy: New Approaches to Scholarship and Teaching from Digital Humanities*, edited by Daniel J. Cohen and Tom Scheinfeldt, 3–5. Ann Arbor: University of Michigan Press. doi:10.3998/dh.12172434.0001.001.

Cohen and Scheinfeldt introduce *Hacking the Academy*, a digital publishing experiment and attempt to reform academic institutions and practices by crowdsourcing content. The editors called for submissions to their project with the caveat that participants had one week to submit. Cohen and Scheinfeldt pitched their project with the following questions: "Can an algorithm edit a journal? Can a library exist without books? Can students build and manage their own learning management platforms? Can a conference be held without a program? Can Twitter replace a scholarly society?" (3). Roughly one sixth of the 329 submissions received were included in the consequent publication. The intent of the project was to reveal the desire and possibility for large institutional change via digital means.

*Fitzpatrick, Kathleen. 2007. "CommentPress: New (Social) Structures for New (Networked) Texts." *Journal of Electronic Publishing* 10 (3): n.p. doi:10.3998/3336451.0010.305.

Fitzpatrick meditates on the current state and future possibilities of electronic scholarly publishing. She focuses her consideration on a study of CommentPress, a digital scholarly publishing venue that combines the hosting of long texts with social network features. Fitzpatrick argues that community and collaboration are at the heart of scholarly knowledge creation—or at least, they should be. Platforms such as CommentPress acknowledge the productive capabilities of scholarly collaboration, and promote this fruitful interaction between academics. Although Fitzpatrick admits that CommentPress is not the only or best answer to the questions of shifting scholarly communication, she celebrates its emergence as a service for the social interconnection and knowledge production of authors and readers in an academic setting.

*_____ 2011. *Planned Obsolescence: Publishing, Technology, and the Future of the Academy*. New York: New York University Press. ("Introduction: Obsolescence," 1–14, and Chapter 3: "Texts," 89–120, are accessible at http://raley.english.ucsb.edu/wp-content2/uploads/234/Fitzpatrick.pdf.)

Fitzpatrick duly surveys and calls for a reform of academic publishing. She argues for more interactivity, communication, and peer-to-peer review, as well as a significant move toward digital scholarly publishing. Fitzpatrick demonstrates that the current mode of scholarly publishing is economically unviable. Moreover, tenure and promotion practices based primarily on traditional modes of scholarly publishing need to be reformed. Fitzpatrick acknowledges certain touchstones of the academy (peer review, scholarship, sharing ideas), and how these tenets have been overshadowed by priorities shaped, in part, by mainstream academic publishing practices and concepts. She details her own work with CommentPress, and the benefits of publishing online with an infrastructure that enables widespread dissemination as well as concurrent reader participation via open peer review.

Huffman, Steve, and Alexis Ohanian. 2005. *Reddit.* **https://www.reddit. com.**

As a popular social news site, Reddit prompts users to tag and submit content. The hierarchy of posts on the front page of the site (as well as the other pages on the site) is decided by a ranking system predicated on both date of submission and voting by other users. Reddit exemplifies social knowledge creation via folksonomy tagging in a social network environment. Notably, the news site is also open source.

*Liu, Alan. 2011. "Friending the Past: The Sense of History and Social Computing." *New Literary History: A Journal of Theory and Interpretation* **42 (1): 1–30. doi:10.1353/nlh.2011.0004.**

Liu identifies media-induced sociality in oral, written, and digital culture. He proceeds to analyze Web 2.0 and social computing practices, and concludes that Web 2.0 lacks a sense of history, despite its intricately interconnected state. Liu attributes this state to two concurrent historical shifts: a social move from one-to-many to many-to-many, and a temporal shift from straightforward conceptions of time into the contemporary conception of instantaneous and simultaneous temporality. Reflexively, Liu argues that conceiving of time in this new instantaneous/simultaneous framework may ideologically proprietize the Internet and allow for ownership of social practices by organizations such as Facebook, Twitter, and Google. As such, Liu opts for a more traditional sense of temporality and history characterized by narratological linear time. He cites the social network system of his Research-oriented Social Environment (RoSE) project as a platform that integrates history with Web 2.0 infrastructure and allowances.

Michel, Jean-Baptiste, Yuan Kui Shen, Aviva Presser Aiden, Adrian Veres, Matthew K. Gray, Google Books Team, Joseph P. Pickett, Dale Hoiberg, Dan Clancy, Peter Norvig, Jon Orwant, Steven Pinker, Martin A. Nowak, and Erez Lieberman Aiden. 2011. "Quantitative Analysis of Culture Using Millions of Digitized Books." *Science* **331 (6014): 176–82. doi:10.1126/science.1199644.**

The authors detail some of the processes and findings of Google's NGram viewer and the related field of study, "culturomics." They argue that an analysis of word frequencies in a large corpus of texts brings to light linguistic and therefore cultural trends. Using word frequency and variation as the predominant metric, Michel et al. discuss various social and historical trends. They do not, however, account for the reductionist concept that word frequency in a selected corpus can attest to or represent all of the varying social movements, actors, and contexts that make up a cultural trend.

Mozilla Foundation. 2011. *Open Badges.* **https://openbadges.org.**
Mozilla's Open Badges is an alternative credential-granting system designed for the public recognition of non-conventional learning and success. Broadly articulated as a democratizing service, Open Badges allows various organizations to accredit their participants within a recognizable system. In an era of Massive Open Online Courses (MOOCs) and citizen scholars, Open Badges embodies the ethos of the decentralized network of contemporary learning, accreditation, and social knowledge creation.

*Nowviskie, Bethany. 2012a. "A Digital Boot Camp for Grad Students in the Humanities." *The Chronicle of Higher Education.* **Last modified April 29, 2012. http://chronicle.com/article/A-Digital-Boot-Camp-for-Grad/131665.**
Nowviskie discusses the Praxis program she directs out of the Scholars' Lab at the University of Virginia. She demonstrates how a combined commitment to interdisciplinarity, collaboration, and tacit knowledge is used to effectively train graduate students in contemporary humanities (and especially digital humanities) work. Nowviskie acknowledges the challenges and benefits of blending radically new methods for graduate training with traditional humanities practices and credit systems. Overall, she reiterates the value of training graduate students in an open-ended, community-minded way; in this way, humanities programs can facilitate both graduate and postgraduate school careers.

Open Knowledge Foundation. 2009–12. *AnnotateIt / Annotator.* **http://annotateit.org.**
AnnotateIt is an effective and easy to use system that enables online annotations. A bookmarklet is used to add the JavaScript tool Annotator to any web page; users can then annotate or comment on various elements on the page, and save the annotations to AnnotateIt. This sort of tool readily facilitates social knowledge creation through collaborative annotation. User annotations may contain tags, content created using the Markdown conversion tool, and individual permissions per annotation. Annotator is also easily extensible, allowing for the potential inclusion of more behaviours or features. Of note, the Open Knowledge Foundation has developed many social knowledge creation tools, including BibServer (https://github.com/okfn/bibserver), CKAN (http://ckan.org/), and TEXTUS (http://textusproject.org/)—all of which are annotated in this bibliography.

Roy Rosenzweig Center for History and New **Media (George Mason University). 2007–13.** *Omeka.* **http://omeka.org.**
Omeka is an example of social knowledge creation through user-driven or generated content. An open source content management system, Omeka was designed to display online digital collections of scholarly editions and cultural heritage artifacts. This content management system acts as a collections management tool and an archival digital collection system, allowing for productive scholarly and non-scholarly exhibitions to develop. Omeka includes an extensive list of features aimed at scholars, museum professionals, librarians, archivists, educators, and other enthusiasts. Of note, the Roy Rosenzweig Center also developed the open bibliography initiative Zotero (included in this annotated bibliography).

13. A Complete Alphabetical List of Selections

Aarseth, Espen. 1997. "Introduction." In *Cybertext: Perspectives on Ergodic Literature*, 1–23. Baltimore: Johns Hopkins University Press.Althusser, Louis. 1971. "Ideology and Ideological State Apparatuses (Notes Towards an Investigation)." In *Lenin and Philosophy and Other Essays*, translated by Ben Brewster, 127–86. New York: Monthly Review Press.

Ancient World Mapping Center, Stoa Consortium, and Institute for the Study of the Ancient World. 2000. *Pleiades.* http://pleiades.stoa.org/.

Andersen, Christian Ulrik, and Søren Bro Pold. 2014. "Post-digital Books and Disruptive Literary Machines." *Formules/Revue des Créations Formelles et Littératures à Contraintes* 18: 169–88.

Ang, Ien. 2004. "Who Needs Cultural Research?" In *Cultural Studies and Practical Politics: Theory, Coalition Building, and Social Activism*, edited by Pepi Leystina, 477–83. New York: Blackwell.

Arbuckle, Alyssa, and Alex Christie, with the ETCL, INKE, and MVP Research Groups. 2015. "Intersections Between Social Knowledge Creation and Critical Making." *Scholarly and Research Communication* 6 (3): n.p. http://src-online.ca/index.php/src/article/view/200.

Arbuckle, Alyssa, Constance Crompton, and Aaron Mauro. 2014. Introduction: "Building Partnerships to Transform Scholarly Publishing." *Scholarly and Research Communication* 5 (4): n.p. http://src-online.ca/index.php/src/article/view/195.

Arbuckle, Alyssa, Aaron Mauro, and Lynne Siemens. 2015. Introduction: "From Technical Standards to Research Communities: Implementing New Knowledge Environments Gatherings, Sydney 2014 and Whistler 2015." *Scholarly and Research Communication* 6 (2): n.p. http://src-online.ca/index.php/src/article/view/232.

Avila, Maria, with contributions from Alan Knoerr, Nik Orlando, and Celestina Castillo. 2010. "Community Organizing Practices in Academia: A Model, and Stories of Partnerships." *Journal of Higher Education Outreach and Engagement* 14 (2): 37–63. http://openjournals.libs.uga.edu/index.php/jheoe/article/view/43/38.

Bachelard, Gaston. 1969. *The Poetics of Space.* Translated by Maria Jolas. Boston: Beacon Press.

Bailey, Moya Z. 2011. "All the Digital Humanists Are White, All the Nerds Are Men, But Some of Us Are Brave." *Journal of Digital Humanities* 1 (1): n.p. http://journalofdigitalhumanities.org/1-1/all-the-digital-humanists-are-white-all-the-nerds-are-men-but-some-of-us-are-brave-by-moya-z-bailey/.

Ball, John Clement. 2010. "Definite Article: Graduate Student Publishing, Pedagogy, and the Journal as Training Ground." *Canadian Literature* 204: 160–62.

Balsamo, Anne. 2011. Introduction: "Taking Culture Seriously in the Age of Innovation." In *Designing Culture: The Technological Imagination at Work*, 2–25. Durham, NC: Duke University Press.

Bath, Jon, and Scott Schofield. 2015. "The Digital Book." In *The Cambridge Companion to the History of the Book,* edited by Leslie Howsam, 181–95. Cambridge: Cambridge University Press.

Bazerman, Charles. 1991. "How Natural Philosophers Can Cooperate: The Literary Technology of Coordinated Investigation in Joseph Priestley's *History and Present State of Electricity* (1767)." In *Textual Dynamics of the Professions: Historical and Contemporary Studies of Writing in Professional Communities*, edited by Charles Bazerman and James Paradis, 13–44. Madison: University of Wisconsin Press.

Benkler, Yochai. 2003. "Freedom in the Commons: Towards a Political Economy of Information." *Duke Law Journal* 52 (6): 1245–76. http://scholarship.law.duke.edu/dlj/vol52/iss6/3.

Berkenkotter, Carol, Thomas N. Huckin, and John Ackerman. 1991. "Social Context and Socially Constructed Texts: The Initiation of a Graduate Student into a Writing Research Community." In *Textual Dynamics of the Professions: Historical and Contemporary Studies of Writing in Professional Communities*, edited by Charles Bazerman and James Paradis, 191–215. Madison: University of Wisconsin Press.

Berry, David M. 2011. "The Computational Turn: Thinking About the Digital Humanities." *Culture Machine* 12: n.p. http://www.culturemachine.net/index.php/cm/article/view/440/470.

_____ 2012. "The Social Epistemologies of Software." *Social Epistemology: A Journal of Knowledge, Culture and Policy* 26 (3–4): 379–98. doi:10.1080/02691728.2012.727191.

Besser, Howard. 2004. "The Past, Present, and Future of Digital Libraries." In *A Companion to Digital Humanities*, edited by Susan Schreibman, Raymond G. Siemens, and John Unsworth, 557–75. Oxford: Blackwell.

Biagioli, Mario. 2002. "From Book Censorship to Academic Peer Review." *Emergences* 12 (1): 11–45. doi:10.1080/1045722022000003435.

Bijker, Wiebe E., and John Law. 1992. "General Introduction." In *Shaping Technology/Building Society: Studies in Sociotechnical Change*, edited by Wiebe E. Bijker and John Law, 1–14. Cambridge, MA: MIT Press.

Bolter, Jay David. 2007. "Digital Media and Art: Always Already Complicit?" *Criticism* 49 (1): 107–18. doi:10.1353/crt.2008.0013.

Boot, Peter. 2012. "Literary Evaluation in Online Communities of Writers and Readers." *Scholarly and Research Communication* 3 (2): n.p. http://src-online.ca/index.php/src/article/view/77/90.

Borgman, Christine L. 2007. *Scholarship in the Digital Age: Information, Infrastructure, and the Internet.* Cambridge, MA: MIT Press.

Bourdieu, Pierre. 1993. "The Field of Cultural Production, or: The Economic World Reversed." In *The Field of Cultural Production: Essays on Art and Literature*, edited and translated by Randal Johnson, 29–73. New York: Columbia University Press.

Bowen, William R., Constance Crompton, and Matthew Hiebert. 2014. "Iter Community: Prototyping an Environment for Social Knowledge Creation and Communication." *Scholarly and Research Communication.* 5 (4): n.p. http://src- online.ca/index.php/src/article/view/193/360.

Brant, Claire. 2011. "The Progress of Knowledge in the Regions of Air?: Divisions and Disciplines in Early Ballooning." *Eighteenth-Century Studies* 45 (1): 71–86. doi:10.1353/ecs.2011.0050.

Brooks, Kevin. 2002. "National Culture and the First-Year English Curriculum: A Historical Study of 'Composition' in Canadian Universities." *American Review of Canadian Studies* 32 (4): 673–94. doi:10.1080/02722010209481679.

Brown, David W. 1995. "The Public/Academic Disconnect." In *Higher Education Exchange Annual*, 38–42. Dayton, OH: Kettering Foundation.

Buehl, Jonathan, Tamar Chute, and Anne Fields. 2012. "Training in the Archives: Archival Research as Professional Development." *College Composition and Communication* 64 (2): 274–305.

Burdick, Anne, Johanna Drucker, Peter Lunenfeld, Todd Presner, and Jeffrey Schnapp. 2012. "The Social Life of the Digital Humanities." In *Digital_Humanities,* 73–98. Cambridge, MA: MIT Press.

Burke, Peter. 2000. *A Social History of Knowledge: From Gutenberg to Diderot.* Cambridge: Polity Press.

_____ 2012. *A Social History of Knowledge II: From the Encyclopédie to Wikipedia.* Cambridge: Polity Press.

Cao, Qilin, Yong Lu, Dayong Dong, Zongming Tang, and Yongqiang Li. 2013. "The Roles of Bridging and Bonding in Social Media Communities." *Journal of the American Society for Information Science and Technology* 64 (8): 1671–81. doi:10.1002/asi.22866.

Carletti, Laura, Derek McAuley, Dominic Price, Gabriella Giannachi, and Steve Benford. 2013. "Digital Humanities and Crowdsourcing: An Exploration." *Museums and the Web 2013 Conference.* Portland: Museums and the Web LLC. http://mw2013.museumsandtheweb.com/paper/digital-humanities-and-crowdsourcing-an-exploration-4/.

Carlton, Susan Brown. 1995. "Composition as a Postdisciplinary Formation." *Rhetoric Review* 14 (1): 78–87. doi:10.1080/07350199509389053.

Causer, Tim, and Melissa Terras. 2014. "Crowdsourcing Bentham: Beyond the Traditional Boundaries of Academic History." *International Journal of Humanities and Arts Computing* 8 (1): 46–64. doi:10.3366/ijhac.2014.0119.

Causer, Tim, Justin Tonra, and Valerie Wallace. 2012. "Transcription Maximized; Expense Minimized? Crowdsourcing and Editing *The Collected Works of Jeremy Bentham*." *Digital Scholarship in the Humanities* (formerly *Literary and Linguistic Computing*) 27 (2): 119–37. doi:10.1093/llc/fqs004.

Causer, Tim, and Valerie Wallace. 2012. "Building a Volunteer Community: Results and Findings from *Transcribe Bentham*." *Digital Humanities Quarterly* 6 (2): n.p. http://digitalhumanities.org:8081/dhq/vol/6/2/000125/000125.html.

Chapman, Owen, and Kim Sawchuk. 2015. "Creation-as-Research: Critical Making in Complex Environments." *RACAR: Revue d'art canadienne / Canadian Art Review* 40 (1): 49–52. http://www.jstor.org/stable/24327426.

Chun, Wendy Hui Kyong. 2004. "On Software, or the Persistence of Visual Knowledge." *Grey Room* 18: 26–51. doi:10.1162/1526381043320741.

Christie, Alex, and the INKE and MVP Research Groups. 2014. "Interdisciplinary, Interactive, and Online: Building Open Communication Through Multimodal Scholarly Articles and Monographs." *Scholarly and Research Communication* 5 (4): n.p. http://src-online.ca/index.php/src/article/view/190.

Clement, Tanya. 2011. "Knowledge Representation and Digital Scholarly Editions in Theory and Practice." *Journal of the Text Encoding Initiative* 1: n.p. doi:10.4000/jtei.203.

Cohen, Daniel J. 2008. "Creating Scholarly Tools and Resources for the Digital Ecosystem: Building Connections in the Zotero Project.'" *First Monday* 13 (8): n.p. doi:10.5210/fm.v13i8.2233.

_____ 2012. "The Social Contract of Scholarly Publishing." In *Debates in the Digital Humanities*, edited by Matthew K. Gold, 319–21. Minneapolis: University of Minnesota Press. http://dhdebates.gc.cuny.edu/debates/text/27.

Cohen, Daniel J., and Tom Scheinfeldt. 2013. "Preface." In *Hacking the Academy: New Approaches to Scholarship and Teaching from Digital Humanities*, edited by Daniel J. Cohen and Tom Scheinfeldt, 3–5. Ann Arbor: University of Michigan Press. doi:10.3998/dh.12172434.0001.001.

Crompton, Constance, Raymond G. Siemens, and Alyssa Arbuckle, with the INKE Research Group. 2015. "Enlisting 'Vertues Noble & Excelent': Behavior, Credit, and Knowledge Organization in the Social Edition." *Digital Humanities Quarterly* 9 (2): n.p. http://www.digitalhumanities.org/dhq/vol/9/2/000202/000202.html.

Davidson, Cathy N., and David Theo Goldberg. 2004. "Engaging the Humanities." *Profession*: 42–62. doi:10.1632/074069504X26386.

De Roure, David. 2014. "The Future of Scholarly Communications." *Insights* 27 (3): 233–38. doi:10.1629/2048-7754.171.

Drucker, Johanna. 2009. "From Digital Humanities to Speculative Computing." In *SpecLab: Digital Aesthetics and Projects in Speculative Computing*, 3–18. Chicago: University of Chicago Press.

_____ 2011. "Humanities Approaches to Interface Theory." *Culture Machine* 12: 1–20. http://www.culturemachine.net/index.php/cm/article/viewArticle/434.

_____ 2012. "Humanistic Theory and Digital Scholarship." In *Debates in the Digital Humanities*, edited by Matthew K. Gold, 85–95. Minneapolis: University of Minnesota Press. http://dhdebates.gc.cuny.edu/debates/text/34.

Eagleton, Terry. 2010. "The Rise of English." In *The Norton Anthology of Theory and Criticism*, edited by Vincent B. Leitch, 2140–46. New York: W.W. Norton.

Edwards, Charlie. 2012. "The Digital Humanities and Its Users." In *Debates in the Digital Humanities*, edited by Matthew K. Gold, 213–32. Minnesota: University of Minnesota Press. http://dhdebates.gc.cuny.edu/debates/text/31.

Ellison, Julie. 2008. "The Humanities and the Public Soul." *Antipode* 40 (3): 463–71. doi:10.1111/j.1467-8330.2008.00615.x.

Ellison, Julie, and Timothy Eatman. 2008. *Scholarship in Public: Knowledge Creation and Tenure Policy in the Engaged University*. Syracuse, NY: Imagining America. http://imaginingamerica.org/wp-content/uploads/2011/05/TTI_FINAL.pdf.

Farland, Maria M. 1996. "Academic Professionalism and the New Public Mindedness." *Higher Education Exchange* Annual: 51–57. http://www.unz. org/Pub/HigherEdExchange-1996q1–00051.

Fisher, Caitlin. 2015. "Mentoring Research-Creation: Secrets, Strategies, and Beautiful Failures." *RACAR: Revue d'art canadienne / Canadian Art Review* 40 (1): 46–49. http://www.jstor.org/stable/24327425.

Fitzpatrick, Kathleen. 2007. "CommentPress: New (Social) Structures for New (Networked) Texts." *Journal of Electronic Publishing* 10 (3): n.p. doi:10.3998/3336451.0010.305.

—————— 2009. "Peer-To-Peer Review and the Future of Scholarly Authority." *Cinema Journal* 48 (2): 124–29. doi:10.1353/cj.0.0095.

—————— 2011. *Planned Obsolescence: Publishing, Technology, and the Future of the Academy.* New York: New York University Press. ("Introduction: Obsolescence," 1–14, and Chapter 3: "Texts," 89–120, are accessible at http://raley.english.ucsb.edu/wp-content2/uploads/234/Fitzpatrick. pdf.)

—————— 2012a. "Beyond Metrics: Community Authorization and Open Peer Review." In *Debates in the Digital Humanities,* edited by Matthew K. Gold, 452–59. Minneapolis: University of Minnesota Press. http:// dhdebates.gc.cuny.edu/debates/text/7.

—————— 2012b. "The Humanities, Done Digitally." In *Debates in the Digital Humanities*, edited by Matthew K. Gold, 12–15. Minneapolis: University of Minnesota Press. http://dhdebates.gc.cuny.edu/debates/text/30.

Fjällbrant, Nancy. 1997. "Scholarly Communication—Historical Development and New Possibilities." In *Proceedings of the IATUL Conferences.* West Lafayette, IN: Purdue University Libraries e-Pubs. http://docs.lib. purdue.edu/cgi/viewcontent.cgi?article=1389&context=iatul.

Flanders, Julia. 2005. "Detailism, Digital Texts, and the Problem of Pedantry." *TEXT Technology* 14 (2): 41–70. http://texttechnology.mcmaster.ca/pdf/ vol14_2/flanders14-2.pdf.

—————— 2009. "The Productive Unease of 21st-Century Digital Scholarship." *Digital Humanities Quarterly* 3 (3): n.p. http://www. digitalhumanities.org/dhq/vol/3/3/000055/000055.html.

_____ 2012. "Time, Labor, and 'Alternate Careers' in Digital Humanities Knowledge Work." In *Debates in the Digital Humanities*, edited by Matthew K. Gold, 292–308. Minneapolis: University of Minnesota Press. http://dhdebates.gc.cuny.edu/debates/text/26.

Franklin, Michael J., Donald Kossman, Tim Kraska, Sukriti Ramesh, and Reynold Xin. 2011. "CrowdDB: Answering Queries with Crowdsourcing." In *Proceedings of the 2011 ACM SIGMOD International Conference on Management of Data (SIGMOND/PODS '11)*, 61–72. New York: ACM.

Fraser, Nancy. 1990. "Rethinking the Public Sphere: A Contribution to the Critique of Actually Existing Democracy." *Social Text* (25, 26): 56–80. http://www.jstor.org/stable/466240.

Freeman, Jo. 1972. "The Tyranny of Structurelessness." *The Second Wave* 2 (1): n.p. http://www.jofreeman.com/joreen/tyranny.htm.

Foucault, Michel. 1977. *Discipline and Punish: The Birth of the Prison.* Translated by Alan Sheridan. London: Allen Lane and Penguin Books.

Garson, Marjorie. 2008. "ACUTE: The First Twenty-Five Years, 1957–1982." *English Studies in Canada* 34 (4): 21–43. https://ejournals.library.ualberta.ca/index.php/ESC/article/view/19771/15285.

Gitelman, Lisa. 2006. *Always Already New: Media, History, and the Data of Culture.* Cambridge, MA: MIT Press.

Ghosh, Arpita, Satyen Kale, and Preston McAfee. 2011. "Who Moderates the Moderators? Crowdsourcing Abuse Detection in User-Generated Content." In *Proceedings of the 12th ACM Conference on Electronic Commerce (EC' 11)*, 167–76. New York: ACM.

Graff, Gerald. 1987. *Professing Literature: An Institutional History.* Chicago: University of Chicago Press.

_____ 2003. "Introduction: In the Dark All Eggheads Are Gray." In *Clueless in Academe: How Schooling Obscures the Life of the Mind,* 1–16. New Haven, CT: Yale University Press.

Gregory, Derek. 1994. *Geographical Imaginations.* Oxford: Blackwell.

Guédon, Jean-Claude. 2008. "Digitizing and the Meaning of Knowledge." *Academic Matters* (October–November): 23–26. http://www.academicmatters.ca/assets/AM_SEPT'08.pdf.

Guldi, Jo. 2013. "Reinventing the Academic Journal." In *Hacking the Academy: New Approaches to Scholarship and Teaching from Digital Humanities,* edited by Daniel J. Cohen and Tom Scheinfeldt, 19–24. Ann Arbor: University of Michigan Press. doi:10.3998/dh.12172434.0001.001.

Guldi, Jo, and Cora Johnson-Roberson. 2012. *Paper Machines.* metaLAB @ Harvard. http://papermachines.org/.

Habermas, Jürgen. 1991. "Introduction: Preliminary Demarcation of a Type of Bourgeois Public Sphere." In *The Structural Transformation of the Public Sphere,* translated by Thomas Burger with the assistance of Frederick Lawrence, 1–26. Cambridge, MA: MIT Press.

Haft, Jamie. 2012. "Publicly Engaged Scholarship in the Humanities, Arts, and Design." *Animating Democracy*: 1–15. http://imaginingamerica.org/wp-content/uploads/2015/09/JHaft-Trend-Paper.pdf.

Hart, Jennefer, Charlene Ridley, Faisal Taher, Corina Sas, and Alan J. Dix. 2008. "Exploring the Facebook Experience: A New Approach to Usability." In *Proceedings of the 5th Nordic Conference on Human-Computer Interaction (NordiCHI08),* 471–74. New York: ACM.

Hart, William, and Terry Marsh. 2014. "Social Media Research Foundation." In *Encyclopedia of Social Media and Politics,* edited by Kerric Harvey, 3: 1173–74. Thousand Oaks, CA: Sage.

Hayles, N. Katherine. 2008. *Electronic Literature: New Horizons for the Literary.* Notre Dame, IN: University of Notre Dame Press.

Haraway, Donna. 1990. "A Cyborg Manifesto: Science, Technology, and Socialist Feminism in the Late Twentieth Century." In *Simians, Cyborgs, and Women: The Reinvention of Nature,* 149–81. New York: Routledge.

Heidegger, Martin. 1982. "The Question Concerning Technology." In *The Question Concerning Technology and Other Questions,* translated with an introduction by William Lovitt, 3–35. New York: Harper Perennial.

Hendry, David G., J.R. Jenkins, and Joseph F. McCarthy. 2006. "Collaborative Bibliography." *Information Processing & Management* 42 (3): 805–25. doi:10.1016/j.ipm.2005.05.007.

Holley, Rose. 2010. "Crowdsourcing: How and Why Should Libraries Do It?" *D-Lib Magazine* 16 (3/4): n.p. doi:10.1045/march2010–holley.

Huffman, Steve, and Alexis Ohanian. 2005. *Reddit.* https://www.reddit.com.

Introna, Lucas D., and Helen Nissenbaum. 2000. "Shaping the Web: Why the Politics of Search Engines Matters." *The Information Society* 16 (3): 169–85.

Inversini, Alessandro, Rogan Sage, Nigel Williams, and Dimitrios Buhalis. 2015. "The Social Impact of Events in Social Media Conversation." In *Information and Communication Technologies in Tourism* 2015, edited by Iis Tussyadiah and Alessandro Inversini, 283–94. Lugano, Switzerland: Springer International Publishing.

Ittersum, Martine J. van. 2011. "Knowledge Production in the Dutch Republic: The Household Academy of Hugo Grotius." *Journal of the History of Ideas* 72 (4): 523–48. doi:10.1353/jhi.2011.0033.

Jagodzinski, Cecile M. 2008. "The University Press in North America: A Brief History." *Journal of Scholarly Publishing* 40 (1): 1–20. doi:10.1353/scp.0.0022.

Jankowski, Nicholas W., Andrea Scharnhorst, Clifford Tatum, and Zuotian Tatum. 2013. "Enhancing Scholarly Publications: Developing Hybrid Monographs in the Humanities and Social Sciences." *Scholarly and Research Communication* 4 (1): n.p. http://src-online.ca/index.php/src/article/view/40/123.

Jay, Gregory. 2012. "The Engaged Humanities: Principles and Practices for Public Scholarship and Teaching." *Journal of Community Engagement and Scholarship* 3 (1): 51–63. http://jces.ua.edu/the-engaged-humanities-principles-and-practices-for-public-scholarship-and-teaching/.

Jenstad, Janelle, and Kim McLean-Fiander. n.d. "The *MoEML* Gazetteer of Early Modern London." *The Map of Early Modern London*, edited by Janelle Jenstad. Victoria: University of Victoria. http://mapoflondon.uvic.ca/gazetteer_about.htm.

Jessop, Martyn. 2008. "Digital Visualization as a Scholarly Activity." *Digital Scholarship in the Humanities* (formerly *Literary and Linguistic Computing*) 23 (3): 281–93. doi:10.1093/llc/fqn016.

Johns, Adrian. 1998. *The Nature of the Book: Print and Knowledge in the Making.* Chicago: University of Chicago Press.

Jones, Steven E. 2013a. *The Emergence of the Digital Humanities.* London and New York: Routledge.

_____ 2013b. "Publications." In *The Emergence of the Digital Humanities,* 147–77. London and New York: Routledge.

Kaufer, David S., and Kathleen M. Carley. 1993. "Academia." In *Communication at a Distance: The Influence of Print on Sociocultural Organization and Change,* 341–93. Hillsdale, NJ: Lawrence Earlbaum Associates.

Kingsley, Danny. 2013. "Build It and They Will Come? Support for Open Access in Australia." *Scholarly and Research Communication* 4 (1): n.p. http://src-online.ca/index.php/src/article/viewFile/39/121.

Kirschenbaum, Matthew. 2012a. "Digital Humanities As/Is a Tactical Term." In *Debates in the Digital Humanities,* edited by Matthew K. Gold, 415–28. Minneapolis: University of Minnesota Press. http://dhdebates.gc.cuny.edu/debates/text/48.

_____ 2012b. "What is Digital Humanities and What's It Doing in English Departments?" In *Debates in the Digital Humanities,* edited by Matthew K. Gold, 3–11. Minneapolis: University of Minnesota Press. http://dhdebates.gc.cuny.edu/debates/text/38.

Kittur, Aniket, and Robert E. Kraut. 2008. "Harnessing the Wisdom of the Crowds in Wikipedia: Quality Through Coordination." In *Proceedings of the 2008 ACM Conference on Computer Supported Cooperative Work (CSCW 08),* 37–46. New York: ACM.

Kjellberg, Sara. 2010. "I am a Blogging Researcher: Motivations for Blogging in a Scholarly Context." *First Monday* 15 (8): n.p. doi:10.5210/fm.v15i8.2962.

Lane, Richard J. 2014. "Innovation through Tradition: New Scholarly Publishing Applications Modelled on Faith-Based Electronic Publishing and Learning Environments." *Scholarly and Research Communication* 5 (4): n.p. http://src-online.ca/index.php/src/article/view/188.

Latour, Bruno. 2009. "A Cautious Prometheus? A Few Steps Towards a Philosophy of Design (with Special Attention to Peter Sloterdijk)." In *Networks of Design: Proceedings of the 2008 Annual International Conference of the Design History Society,* edited by Fiona Hackne, Jonathan Glynne, and Viv Minto, 2–10. Boca Raton, FL: Universal Publishers.

Lessig, Lawrence. 2004. *Free Culture: How Big Media Uses Technology and the Law to Lock Down Culture and Control Creativity*. New York: Penguin.

Lightman, Harriet, and Ruth N. Reingold. 2005. "A Collaborative Model for Teaching E-Resources: Northwestern University's Graduate Training Day." *Libraries and the Academy* 5 (1): 23–32. doi:10.1353/pla.2005.0008.

Liu, Alan. 2011. "Friending the Past: The Sense of History and Social Computing." *New Literary History: A Journal of Theory and Interpretation* 42 (1): 1–30. doi:10.1353/nlh.2011.0004.

_____ 2013. "From Reading to Social Computing." In *Literary Studies in the Digital Age: An Evolving Anthology*, edited by Kenneth M. Price and Raymond G. Siemens, n.p. New York: MLA Commons. https://dlsanthology.commons.mla.org/from-reading-to-social-computing/.

_____ 2009. "The End of the End of the Book: Dead Books, Lively Margins, and Social Computing." *Michigan Quarterly Review* 48 (4): 499–520.

_____ 2004. *The Laws of Cool: Knowledge Work and the Culture of Information*. Chicago: University of Chicago Press.

_____ 2012. "Where is Cultural Criticism in the Digital Humanities?" In *Debates in the Digital Humanities*, edited by Matthew K. Gold, 490–510. Minneapolis: University of Minnesota Press. http://dhdebates.gc.cuny.edu/debates/text/20.

Lorimer, Rowland. 2013. "Libraries, Scholars, and Publishers in Digital Journal and Monograph Publishing." *Scholarly and Research Communication* 4 (1): n.p. http://src-online.ca/index.php/src/article/viewFile/43/117.

_____ 2014. "A Good Idea, a Difficult Reality: Toward a Publisher/Library Open Access Partnership." *Scholarly and Research Communication* 5 (4): n.p. n.p. http://src-online.ca/index.php/src/article/view/180.

Losh, Elizabeth. 2012. "Hacktivism and the Humanities: Programming Protest in the Era of the Digital University." In *Debates in the Digital Humanities*, edited by Matthew K. Gold, 161–86. Minneapolis: University of Minnesota Press. http://dhdebates.gc.cuny.edu/debates/text/32.

Manovich, Lev. 2001. *The Language of New Media*. Cambridge, MA: MIT Press.

_____ 2012. "Trending: The Promises and the Challenges of Big Social Data." In *Debates in the Digital Humanities*, edited by Matthew K. Gold, 460–75. Minneapolis: University of Minnesota Press. http://dhdebates. gc.cuny.edu/debates/text/15.

Manzo, Christina, Geoff Kaufman, Sukdith Punjasthitkul, and Mary Flanagan. 2015. "'By the People, For the People': Assessing the Value of Crowd-sourced, User-Generated Metadata." *Digital Humanities Quarterly* 9 (1): n.p. http://www.digitalhumanities.org/dhq/vol/9/1/000204/000204. html.

Maxwell, John W. 2014. "Publishing Education in the 21st Century and the Role of the University." *Journal of Electronic Publishing* 17 (2): n.p. http:// quod.lib.umich.edu/j/jep/3336451.0017.205?view=text;rgn=main.

_____ 2015. "Beyond Open Access to Open Publication and Open Scholarship." *Scholarly and Research Communication* 6 (3): n.p. http://src-online.ca/index.php/src/article/view/202.

McCarty, Willard. 2005. *Humanities Computing.* New York: Palgrave Macmillan.

McGann, Jerome. 2006. "From Text to Work: Digital Tools and the Emergence of the Social Text." *TEXT: An Interdisciplinary Annual of Textual Studies* 49–62. http://www.jstor.org/stable/30227956.

_____ 1991. *The Textual Condition.* Princeton, NJ: Princeton University Press.

McGillivray, David, Gayle McPherson, Jennifer Jones, and Alison McCandlish. 2016. "Young People, Digital Media Making and Critical Digital Citizenship." *Leisure Studies* 35 (6): 724–38. doi:10.1080/02614367.2015.1 062041.

McGregor, Heidi, and Kevin Guthrie. 2015. "Delivering Impact of Scholarly Information: Is Access Enough?" *Journal of Electronic Publishing* 18 (3): n.p. http://quod.lib.umich.edu/j/jep/3336451.0018.302?view=text;rgn=main.

McKinley, Donelle. 2012. "Practical Management Strategies for Crowdsourcing in Libraries, Archives and Museums." Report for the School of Information Management, Faculty of Commerce and Administration, Victoria University of Wellington (New Zealand): n.p. http://nonprofitcrowd. org/wp-content/uploads/2014/11/McKinley-2012-Crowdsourcing-management-strategies.pdf.

McPherson, Tara. 2012. "Why are the Digital Humanities So White? or Thinking the Histories of Race and Computation." In *Debates in the Digital Humanities*, edited by Matthew K. Gold, 139–60. Minnesota: University of Minnesota Press. http://dhdebates.gc.cuny.edu/debates/text/29.

Meadows, Alice. 2015. "Beyond Open: Expanding Access to Scholarly Content." *Journal of Electronic Publishing* 18 (3): n.p. http://quod.lib.umich.edu/j/jep/3336451.0018.301?view=text;rgn=main.

Michel, Jean-Baptiste, Yuan Kui Shen, Aviva Presser Aiden, Adrian Veres, Matthew K. Gray, Google Books Team, Joseph P. Pickett, Dale Hoiberg, Dan Clancy, Peter Norvig, Jon Orwant, Steven Pinker, Martin A. Nowak, and Erez Lieberman Aiden. 2011. "Quantitative Analysis of Culture Using Millions of Digitized Books." *Science* 331 (6014): 176–82. doi:10.1126/science.1199644.

Moretti, Franco. 1998. *Atlas of the European Novel, 1800–1900.* London: Verso.

_____ 2005. *Graphs, Maps, Trees: Abstract Models for a Literary History.* London and New York: Verso.

Moyle, Martin, Justin Tonra, and Valerie Wallace. 2011. "Manuscript Transcription by Crowdsourcing: Transcribe Bentham." *Liber Quarterly* 20 (3–4): 347–56. doi:10.18352/lq.7999.

Mozilla Foundation. 2011. *Open Badges.* https://openbadges.org.

Mrva-Montoya, Agata. 2012. "Social Media: New Editing Tools or Weapons of Mass Distraction?" *Journal of Electronic Publishing* 15 (1): 1–24. doi:10.3998/3336451.0015.103.

Nowviskie, Bethany. 2012a. "A Digital Boot Camp for Grad Students in the Humanities." *The Chronicle of Higher Education.* Last modified April 29, 2012. http://chronicle.com/article/A-Digital-Boot-Camp-for-Grad/131665.

_____ 2012b. "Evaluating Collaborative Digital Scholarship (or, Where Credit is Due)." *Journal of Digital Humanities* 1 (4): n.p. http://journalofdigitalhumanities.org/1-4/evaluating-collaborative-digital-scholarship-by-bethany-nowviskie/.

O'Donnell, Daniel, Heather Hobma, Sandra Cowan, Gillian Ayers, Jessica Bay, Marinus Swanepoel, Wendy Merkley, Kelaine Devine, Emma Dering, and Inge Genee. 2015. "Aligning Open Access Publication with the

Research and Teaching Missions of the Public University: The Case of the Lethbridge Journal Incubator (If 'if's and 'and's were pots and pans)." *Journal of Electronic Publishing* 18 (3): n.p. doi:10.3998/3336451.0018.309. http://quod.lib.umich.edu/j/jep/3336451.0018.309?view=text;rgn=main.

Open Knowledge Foundation. 2009–12. *AnnotateIt / Annotator.* http://annotateit.org.

OpenStreetMap Foundation. n.d. *OpenStreetMap.* https://www.openstreetmap.org.

Pearce, Nick, Martin Weller, Eileen Scanlon, and Sam Kinsley. 2010. "Digital Scholarship Considered: How New Technologies Could Transform Academic Work." *Education* 16 (1): n.p. http://ineducation.ca/ineducation/article/view/44.

Pfister, Damien Smith. 2011. "Networked Expertise in the Era of Many-to-Many Communication: On Wikipedia and Invention." *Social Epistemology: A Journal of Knowledge, Culture and Policy* 25 (3): 217–31. doi:10.1080/0269 1728.2011.578306.

Powell, Daniel, Raymond G. Siemens, and William R. Bowen, with Matthew Hiebert and Lindsey Seatter. 2015. "Transformation through Integration: The Renaissance Knowledge Network (ReKN) and a Next Wave of Scholarly Publication." *Scholarly and Research Communication* 6 (2): n.p. http://src-online.ca/index.php/src/article/view/199.

Powell, Daniel, Raymond G. Siemens, and the INKE Research Group. 2014. "Building Alternative Scholarly Publishing Capacity: The Renaissance Knowledge Network (ReKN) as Digital Production Hub." *Scholarly and Research Communication* 5 (4): n.p. http://src-online.ca/index.php/src/article/view/183.

Ramsay, Stephen, and Geoffrey Rockwell. 2012. "Developing Things: Notes Toward an Epistemology of Building in the Digital Humanities." In *Debates in the Digital Humanities,* edited by Matthew K. Gold, 75–84. Minneapolis: University of Minnesota Press.

Ratto, Matt. 2011a. "Critical Making: Conceptual and Material Studies in Technology and Social Life." *The Information Society* 27 (4): 252–60. doi:10 .1080/01972243.2011.583819.

_____ 2011b. "Open Design and Critical Making." In *Open Design Now: Why Design Cannot Remain Exclusive*, edited by Bas van Abel, Lucas Evers, Roel Klaassen, and Peter Troxler, n.p. Amsterdam: BIS Publishers. http:// opendesignnow.org/index.php/article/critical-making-matt-ratto/.

Ratto, Matt, and Robert Ree. 2012. "Materializing Information: 3D Printing and Social Change." *First Monday* 17 (7): n.p. doi:10.5210/fm.v17i7.3968.

Ratto, Matt, Sara Ann Wylie, and Kirk Jalbert. 2014. "Introduction to the Special Forum on Critical Making as Research Program." *The Information Society* 30 (2): 85–95. doi:10.1080/01972243.2014.875767.

Ridge, Mia. 2013. "From Tagging to Theorizing: Deepening Engagement with Cultural Heritage through Crowdsourcing." *Curator: The Museum Journal* 56 (4): 435–50. doi:10.1111/cura.12046.

Robinson, Peter. 2010. "Electronic Editions for Everyone." In *Text and Genre in Reconstruction: Effects of Digitalization on Ideas, Behaviours, Products and Institutions*, edited by Willard McCarty, 145–63. Cambridge: Open Book Publishers.

Rockwell, Geoffrey. 2012. "Crowdsourcing the Humanities: Social Research and Collaboration." In *Collaborative Research in the Digital Humanities*, edited by Marilyn Deegan and Willard McCarty, 135–54. Farnham, UK, and Burlington, VT: Ashgate.

Rosenzweig, Roy. 2006. "Can History Be Open Source? Wikipedia and the Future of the Past." *Journal of American History* 93 (1): 117–46.

Ross, Anthony, and Nadia Caidi. 2005. "Action and Reaction: Libraries in the Post 9/11 Environment." *Library and Information Science Research* 27 (1): 97–114. doi:10.1016/j.lisr.2004.09.006.

Ross, Stephen, Alex Christie, and Jentery Sayers. 2014. "Expert/Crowdsourcing for the Linked Modernisms Project." *Scholarly and Research Communication* 5 (4): n.p. http://src-online.ca/index.php/src/article/viewFile/186/368.

Roy Rosenzweig Center for History and New Media (George Mason University). 2007–13. *Omeka*. http://omeka.org.

Saklofske, Jon, with the INKE Research Group. 2012. "Fluid Layering: Reimagining Digital Literary Archives Through Dynamic, User-Generated

Content." *Scholarly and Research Communication* 3 (4): n.p. http://src-online.ca/index.php/src/article/viewFile/70/181.

Saklofske, Jon, and Jake Bruce, with the INKE Modelling and Prototyping Team and the INKE Team. 2013. "Beyond Browsing and Reading: The Open Work of Digital Scholarly Editions." *Scholarly and Research Communication* 4 (3): n.p. http://src-online.ca/index.php/src/article/view/119.

San Martin, Patricia Silvana, Paola Caroline Bongiovani, Ana Casali, and Claudia Deco. 2015. "Study on Perspectives Regarding Deposit on Open Access Repositories in the Context of Public Universities in the Central-Eastern Region of Argentina." *Scholarly and Research Communication* 6 (1): n.p. http://src-online.ca/index.php/src/article/view/145.

Shillingsburg, Peter. 2006. *From Gutenberg to Google: Electronic Representations of Literary Texts.* Cambridge: Cambridge University Press.

Siemens, Lynne. 2009. "It's a Team if You Use 'Reply All': An Exploration of Research Teams in Digital Humanities Environments." *Digital Scholarship in the Humanities* (formerly *Literary and Linguistic Computing*) 24 (2): 225–33. doi:10.1093/llc/fqp009.

Siemens, Raymond G. 2002. "Scholarly Publishing at its Source, and at Present." In *The Credibility of Electronic Publishing: A Report to the Humanities and Social Sciences Federation of Canada*, compiled by Raymond G. Siemens, Michael Best, Elizabeth Grove-White, Alan Burk, James Kerr, Andy Pope, Jean-Claude Guédon, Geoffrey Rockwell, and Lynne Siemens. *TEXT Technology* 11 (1): 1–128.

Siemens, Raymond G., Meagan Timney, Cara Leitch, Corina Koolen, and Alex Garnett, with the ETCL, INKE, and PKP Research Groups. 2012. "Toward Modeling the *Social* Edition: An Approach to Understanding the Electronic Scholarly Edition in the Context of New and Emerging Social Media." *Digital Scholarship in the Humanities* (formerly *Literary and Linguistic Computing*) 27 (4): 445–61. doi:10.1093/llc/fqs013.

Smith, Martha Nell. 2004. "Electronic Scholarly Editing." In *A Companion to Digital Humanities*, edited by Susan Schreibman, Raymond G. Siemens, and John Unsworth, 306–22. Oxford: Blackwell.

Stanford Natural Language Processing Group. 2006. *Stanford Named Entity Recognizer (NER).* http://nlp.stanford.edu/software/CRF-NER.html.

Stein, Bob. 2015. "Back to the Future." *Journal of Electronic Publishing* 18 (2): n.p. doi:10.3998/3336451.0018.204.

Streeter, Thomas. 2010. "Introduction." In *The Net Effect: Romanticism, Capitalism, and the Internet*, 1–16. New York and London: New York University Press.

Svensson, Patrik. 2012. "Beyond the Big Tent." In *Debates in the Digital Humanities*, edited by Matthew K. Gold, 36–49. Minneapolis: University of Minnesota Press. http://dhdebates.gc.cuny.edu/debates/text/22.

Tally, Robert T., Jr. 2013. *Spatiality*. London and New York: Routledge.

Turner, Fred. 2006. *From Counterculture to Cyberculture: Stewart Brand, the Whole Earth Network, and the Rise of Digital Utopianism.* Chicago: University of Chicago Press.

Vaidhyanathan, Siva. 2002. "The Content-Provider Paradox: Universities in the Information Ecosystem." *Academe* 88 (5): 34–37. doi:10.2307/40252219.

Vandendorpe, Christian. 2015. "Wikipedia and the Ecosystem of Knowledge." *Scholarly and Research Communication* 6 (3): n.p. http://src-online.ca/index.php/src/article/view/201.

_____ 2012. "Wikisource and the Scholarly Book." *Scholarly and Research Communication* 3 (4): n.p. http://src-online.ca/src/index.php/src/article/viewFile/58/146.

Van de Sompel, Herbert, Sandy Payette, John Erickson, Carl Lagoze, and Simeon Warner. 2004. "Rethinking Scholarly Communication: Building the System that Scholars Deserve." *D-Lib Magazine* 10 (9): n.p. doi:10.1045/september2004-vandesompel. http://www.dlib.org/dlib/september04/vandesompel/09vandesompel.html.

Van House, Nancy A. 2003. "Digital Libraries and Collaborative Knowledge Construction." In *Digital Library Use: Social Practice in Design and Evaluation*, edited by Ann Peterson Bishop, Nancy A. Van House, and Barbara P. Buttenfield, 271–95. Cambridge, MA: MIT Press.

Vetch, Paul. 2010. "From Edition to Experience: Feeling the Way towards User-Focused Interfaces." In *Electronic Publishing: Politics and Pragmatics*, edited by Gabriel Egan, 171–84. New Technologies in Medieval and

Renaissance Studies 2. Tempe, AZ: Iter Inc., in collaboration with the Arizona Center for Medieval and Renaissance Studies.

Walsh, Brandon, Claire Maiers, Gwen Nally, Jeremy Boggs, and Praxis Program Team. 2014. "Crowdsourcing Individual Interpretations: Between Microtasking and Multitasking." *Digital Scholarship in the Humanities* (formerly *Literary and Linguistic Computing*) 29 (3): 379–86. doi:10.1093/llc/fqu030.

Wasik, Bill. 2009. *And Then There's This: How Stories Live and Die in Viral Culture.* New York: Viking.

Westphal, Bertrand. 2011. *Geocriticism: Real and Fictional Spaces.* Translated by Robert T. Tally, Jr. New York: Palgrave Macmillan.

Wick, Marc (founder), and Christophe Boutreux (developer). *GeoNames.* Männedorf, Switzerland: Unxos GmbH. http://www.geonames.org.

Williams, George H. 2012. "Disability, Universal Design, and the Digital Humanities." In *Debates in the Digital Humanities*, edited by Matthew K. Gold, 202–12. Minneapolis: University of Minnesota Press. http://dhdebates.gc.cuny.edu/debates/text/44.

Wrisley, David J., and the team at the American University of Beirut. 2016. *Linguistic Landscapes of Beirut* (formerly *Mapping Language Contact in Beirut*). http://llb.djwrisley.com/.

Wylie, Sara Ann, Kirk Jalbert, Shannon Dosemagen, and Matt Ratto. 2014. "Institutions for Civic Technoscience: How Critical Making is Transforming Environmental Research." *The Information Society* 30 (2): 116–26. doi:10.1080/01972243.2014.875767.

Zacharias, Robert. 2011. "The Death of the Graduate Student (and the Birth of the HQP)." *English Studies in Canada* 37 (1): 4–8. https://ejournals.library.ualberta.ca/index.php/ESC/article/view/25197/18694.

II. Game-Design Models for Digital Social Knowledge Creation

Social knowledge creation has the potential to grow and flourish in the Web 2.0 environment through social networking models, crowdsourcing, folksonomic tagging systems, collaborative writing platforms, cloud-based computing, and a variety of many-to-many communication methods. Similarly, video games continue to evolve in exciting ways, especially in the context of ubiquitous networked computers, smartphones, and tablets. Although game studies, as a field, is discussed widely, the ways in which digital scholarship and game studies overlap and relate to each other are not always clear. As digital humanities practices (including multimodal communication, collaborative writing, modelling and prototyping, and hands-on making) become more widespread, the possibility for sharing lessons and insights between game studies and digital humanities increases. Certain scholars may be skeptical of such intersections, but game-based pedagogy projects and humanities-related serious games indicate that overlap has already taken place. This section of the annotated bibliography includes a selection of texts on game-design models, as well as definitions, discourses, and best practices relevant to social knowledge creation.

The application of game-based models in digital scholarship is unsurprising. Games are known for their potential to capture the player's attention, encourage focus and concentration, facilitate collaboration among large groups, and express complex stories and topics in intuitive, experiential ways. As digital humanities practitioners, in particular, develop scholarly and pedagogical environments, these benefits will become increasingly valuable. Perhaps the most widely known game-design approach that is applied in non-game environments is gamification. Gamification falls into a peculiar position within the game-studies/digital humanities relationship: its genesis in the gaming world positions it in the realm of game studies, but its application and definitions necessarily diversify this position. While the term is often used in an ambiguous sense, referring to all game-like or gaming-inspired instances in non-gaming contexts, many scholars differentiate justly between gamification, serious games, playful design, and other related approaches. Sebastian Deterding et al. (2011) offer a well-articulated definition, stating that gamification is "the use of game design elements in non-game contexts" (9), but they also note that gameful design may be a better term for use within academic contexts, since it comes with less baggage than gamification (14). In addition to the negative connotations associated with gamification, the particular focus on implementable game mechanics and elements may limit the potential of the approach. For this reason we use the terms *gameful design, game-design models, game-design thinking,* or *game-inspired approaches* to refer to the broader potential of applying such methods in the development of non-game environments. We hope that this approach resists the reduction of game-design to common game elements and instead aims to apply broader game-design practices and approaches in the development of non-game environments.

Humanities scholars often eschew game-design approaches because of the exploitative reputation of gamification, which is particularly popular in corporate and consumer-facing environments that aim to increase user engagement with a site, program, or application. Because games are so effective at capturing attention and driving engagement, companies and organizations can apply gamification methods to encourage forms of free, immaterial labour from users, and to apply covert methods of driving profits and success rates. In this way, gamification is a prime example of the blurring between play and labour that critics such as Ian Bogost, Alexander Galloway, Trebor Scholz, McKenzie Wark, and Nick Dyer-Witheford and Greig de Peuter study. However, rather than assuming that all game-design-inspired approaches are exploitative across all contexts, this section aims to open up the discourse to acknowledge and engage with critiques of socioeconomic and academic structures. Concurrently, this bibliography draws attention to innovative and practice-based texts on game studies and game design that may inspire scholars to develop game-based responses and solutions.

While certain game-design applications in non-game environments may seem reductive, we believe that a game-inspired design approach can, in fact, help to design sophisticated, self-reflexive environments. These environments benefit from the iterative prototyping process of game design, and also apply procedural rhetoric and effective game mechanics in order to communicate complex arguments in practice. In a social knowledge creation context, game-design models are still in their early stages, and scholarly work on the topic is scarce. As such, the selections below focus on specific areas that aim to offer insights into the critical discourse regarding socioeconomic and institutional practices related to game-design models and social knowledge creation. Ideally, the selections will inspire interested scholars and practitioners to use game-design methods to overcome challenges in social knowledge creation environments. We recommend that readers approach the selections in this section with the above-mentioned vision of game-design-inspired thinking in mind, and consider its potential in the design of social knowledge creation tools and environments. While a number of texts listed below do not discuss game-design methods directly, they cover important issues, concepts, and theories that offer relevant considerations for practitioners who plan to study or implement game-design approaches.

This section of the bibliography consists primarily of sources from the past decade, although a few exceptions were made for particularly relevant texts. The majority of the selections fall under the purview of scholarly, humanities-related work; we have, however, included selections from other areas to reflect accurately the interdisciplinary nature of the proposed game-design inspired practice. Our intention is to provide scholars and practitioners with a present-day survey of popular, widely studied game-design practices while offering a snapshot of discourses and concerns regarding academic humanities practices, video games and game-design studies, and related aspects of the digital landscape and economy. Examples of relevant video games, social networks, and applications also make up a portion of the bibliography. Rather than attempting to cover all relevant video games and applications or offer a history of video games, we include select examples that are either referenced widely, offer particular insight into the origins and practices of game-design applications in non-game contexts, provide inspiring examples from the indie game development movement, or provide a unique, stimulating indication of how games can be applied for scholarly or pedagogical purposes. Additionally, a small number of texts from other industries warranted inclusion based on reception and topical relevance (see Zichermann and Cunningham). The bibliography has been organized into the following six sections consisting of 105 entries, followed by a complete alphabetical list:[9]

1. Game-Design Models in Scholarly Communication Practices and Digital Scholarship

2. Game-Design-Inspired Learning Initiatives

3. Game-Design Models in the Context of Social Knowledge Creation Tools

4. Defining Gamification and Other Game-Design Models

---

[9] Please note that cross-posted entries are marked with an asterisk (*) after the first instance.

5. Game-Design Models and the Digital Economy

6. Game-Design Insights and Best Practices

7. A Complete Alphabetical List of Selections

The initial categories provide a basis for scholarly practices and challenges concerning social knowledge creation. Scholarly communication is an evolving and much-debated activity, and related discourse ranges from issues of tenure track, peer review, and engagement in the digital humanities to the ways in which knowledge and history are presented via Web 2.0 practices and the opportunities social data collection heralds for initiating change. Based on current changes in and criticism of scholarly communication practices and digital scholarship, "Game-Design Models in Scholarly Communication Practices and Digital Scholarship" considers how game-design-inspired engagement, task definition, goal orientation, and collaboration practices can offer new ways of tackling the changes taking place in the humanities. Within the realm of digital scholarship, scholars have begun to consider digital editions as unique spaces for gameful design to be applied. Rather than, for instance, suggesting the simple placement of game-design elements—such as points systems or badges—into a social edition environment, the 22 sources in this section offer critical and conceptual considerations for approaching social knowledge creation from a game-design perspective. The 10 entries in the "Game-Design-Inspired Learning Initiatives" category look at different learning spaces in relation to game-design inspired approaches and models from game environments (such as Massively Multiplayer Online Games [MMOGs] and Massively Multiplayer Online Role-Playing Games [MMORPGs]) in order to demonstrate how they can create collaborative, engaging, and goal-oriented interactive learning environments. The instructional potential of and possibility for learning through games is not a new concept in the realm of pedagogy and teaching. Scholars and teachers have long recognized that engaging students in certain gameplay activities can capture attention, encourage focused and strategic thinking, and teach skills and knowledge. Beyond the actual playing of games, however, game-design thinking can also contribute to the structuring of successful learning environments.

The third category, "Game-Design Models in the Context of Social Knowledge Creation Tools," outlines a select overview of gamification and game-related approaches in particular tools and environments. This category contains a sampling of 26 texts on and examples of social knowledge creation tools, social networks, game platforms, game types, and social literary-analysis environments. It aims to offer an overview of applications and practical insights on the potential of game-design models in the development of social knowledge creation tools. Covering an array of environments, the selections indicate not only how gameful design can encourage user engagement and participation, but also the possible interoperable effects of game environments in the context of social knowledge creation. As Johanna Drucker, Steven Jones, Alan Liu, Jerome McGann, and Geoffrey Rockwell indicate, game interfaces can inspire critical awareness, enable learning by doing (or by modelling, as Jones notes), and integrate otherwise disparate components and interactions, thus provoking deeper forms of collaboration.

The fourth category, "Defining Gamification and Other Game-Design Models," contains 10 selections that discuss the much-debated terminology and definitions of gamification and related approaches. A wide range of fields, from marketing to pedagogy to human resources, apply, study, define, and discuss gamification. While Zichermann and Cunningham (2011) offer a fairly broad definition of gamification as "game-thinking and game mechanics to engage users and solve problems" (xiv), Deterding et al. (2011) differentiate gamification from similar approaches by defining it as "the use of game design elements in non-game contexts" (9). For the purpose of specificity in the context of this bibliography, we follow Deterding's definition and use "gameful design," "game-design thinking," and "game-inspired approaches" to refer to our suggested broader use of game-related methods and strategies in non-game environments. The definitions in this category and their relation to similar approaches provoke debates about terminology, especially because the word "gamification" holds negative connotations associated with marketing tactics. Many scholars, including Deterding and Bogost, argue for alternative terminology in order to distance academic uses of gamification from controversial or exploitative examples.

"Game-Design Models and the Digital Economy" discusses certain key concerns and risks associated with current socioeconomic structures and cultural habits. Within academic discourse, gamification has provoked heated debates and strong criticism. This is not surprising, as video games, and particularly the objectives of gamification, epitomize the play/labour dichotomy. The 15 texts in this category offer varying views of the digital economy with the aim of engendering critical approaches to potential implementations of gamification. While some scholars are highly skeptical of gamification, we believe that game-design models can be used in an ethical and transparent manner. Rather than applying game approaches in an exploitative manner, we see the potential for game-inspired design practices to offer methods that encourage self-reflexivity, critical thinking, and creative engagement. The digital economy in general, and video games in particular, often face challenges as to how to engage scholars and the public in an ethical manner—especially concerning the blurring boundaries between labour and play, entertainment and payment. Furthermore, social shifts in the value and forms of attention are taking place (see Jonathan Beller 2006, N. Katherine Hayles 2007), and the study of game environments is being reformulated and problematized by approaches such as object-oriented ontology and procedural rhetoric (Bogost 2012). Taking these discourses into consideration, the challenge will be to develop uses of gameful design that not only overcome these issues, but contain responses and solutions to them.

Building on the critical base of the previous sections, the final focus on "Game-Design Insights and Best Practices" consists of a selection of game-design-related approaches and practices intended to inform the more practical requirements of developing social knowledge creation tools and environments that incorporate game-design-inspired approaches. The 16 selections cover game-design approaches, best practices, models, and how-tos. Salen and Zimmerman's *Rules of Play,* Bjork and Jussi's *Patterns in Game Design,* and Galloway's *Gaming* offer extensive overviews of video game studies and game design, providing insights into practices from game studies and the gaming industry. The selections in this section of the annotated bibliography, *Game-Design Models for Digital Social Knowledge Creation,* aspire to provide a broad overview of examples, instructions, and approaches to inform practitioners of the possibilities of game-design thinking in social knowledge creation tools and environments, and to assert that game-design-inspired approaches have the potential to offer critical responses and solutions, if applied conscientiously.

1. Game-Design Models in Scholarly Communication Practices and Digital Scholarship

*Aarseth, Espen J. 1997. "Introduction." In *Cybertext: Perspectives on Ergodic Literature,* 1–23. Baltimore: Johns Hopkins University Press.
Aarseth attempts to develop a theory of cybertext works, with a focus on "ergodic texts." Aarseth's scholarly interest lies in texts that are purposefully shaped by the reader's tangible and visible actions and decisions. He bases his speculation on the concept that cybertexts are labyrinthine and user-dependent, and contain feedback loops. Aarseth criticizes the counterarguments that many texts can be read as cybertexts, but does not concede that this distinction derives from cybertexts' necessarily electronic mode.

The inherent performativity involved in reading cybertexts occurs in a network of various parts and participants, compared to the more conventional reading model of reader/author/text. Further, Aarseth argues, ergodic texts (primarily virtual games and MUDs) are defined by the agency and authority of the human subject (reader) whose decisions affect the outcome of the text as a whole.

*Balsamo, Anne. 2011. Introduction: "Taking Culture Seriously in the Age of Innovation." In *Designing Culture: The Technological Imagination at Work*, 2–25. Durham, NC: Duke University Press.

Balsamo studies the intersections of culture and innovation and acknowledges the unity between the two modes ("technoculture"). She argues that technological innovation should seriously recognize culture as both its inherent context and a space of evolving, emergent possibility, as innovation necessarily alters culture and social knowledge creation practices. Balsamo introduces the concept of the "technological imagination"—the innovative, actualizing mindset. She also details a comprehensive list of truisms about technological innovation, ranging from considering innovation as performative, historically constituted, and multidisciplinary to acknowledging design as a major player in cultural reproduction, social negotiation, and meaning-making. Currently, innovation is firmly bound up with economic incentives, and the profit-driven mentality often obscures the social and cultural consequences and implications of technological advancement. As such, Balsamo calls for more conscientious design, education, and development of technology, and a broader vision of the widespread influence and agency of innovation.

Chamberlin, Barbara, Jesús Trespalacios, and Rachel Gallagher. 2014. "Bridging Research and Game Development: A Learning Games Design Model for Multi-Game Projects." In *Educational Technology Use and Design for Improved Learning Opportunities*, edited by Mehdi Khosrow-Pour, 151–71. Hershey, PA: IGI Global. doi:10.4018/978-1-4666-6102-8.ch00.

Chamberlin, Trespalacios, and Gallagher describe the learning games design model for multi-game projects as used to develop math games. The authors present a design approach that integrates content, instructional design, and gaming aspects via a complete project team. They continue to discuss the implications of their approach for other educational games. Although research on game development is available, the authors say that research on educational game development is still relatively new and rare. They situate their study through the affirmation of adapting game play in learning. The authors conclude that development teams may refer to the following critical components for production of new projects: immersion, guiding questions, and team formation.

*Clement, Tanya. 2011. "Knowledge Representation and Digital Scholarly Editions in Theory and Practice." *Journal of the Text Encoding Initiative* **1: n.p. doi:10.4000/jtei.203.**

Clement reflects on scholarly digital editions as sites of textual performance, wherein the editor lays down and privileges various narrative threads for the reader to pick up and interpret. She underscores this theoretical discussion with examples from her own work with the digital edition *In Transition: Selected Poems by the Baroness Elsa von Freytag-Loringhoven*, as well as TEI and XML encoding and the Versioning Machine. Clement details how editorial decisions shape the social experience of an edition. By applying John Bryant's theory of the fluid text to her own editorial practice, she focuses on concepts of various textual performances and meaning-making events. Notably, Clement also explores the idea of the social text network. She concludes that the concept of the network is not new to digital editions; nevertheless, conceiving of a digital edition as a network of various players, temporal spaces, and instantiations promotes fruitful scholarly exploration.

Davidson, Cathy N. 2011. "Why Badges? Why Not?" [blog post]. *HASTAC.* **https://www.hastac.org/blogs/cathy-davidson/2011/09/16/why-badges-why-not.**

In this much-debated *HASTAC* post, Davidson argues in support of the "Badges for Lifelong Learning" competition and for the use of badges as an alternate credential system in academia, training, and education. She notes that one of the key benefits of badges is that they recognize achievement and contribution over reputation or credentials, and thus offer alternatives to current institutional and educational credentials and evaluation standards. This blog post incited an extensive discussion about badges as a new credential system. In the comments section, Ian Bogost offers a critical view, pointing out issues such as the false dichotomy between badges and the current letter-grade system, the question of standardization of badges, and issues such as the labour metrics that go with badge systems.

*Davidson, Cathy N., and David Theo Goldberg. 2004. "Engaging the Humanities." *Profession:* **42–62. doi:10.1632/074069504X26386.**

Davidson and Goldberg argue that despite marginalization, humanistic approaches and perspectives remain significant for successful, holistic university environments. Rather than taking a field-specific approach, Davidson and Goldberg propose a problem- or issue-based humanities model that allows for a more interdisciplinary approach. In this way, the comprehensive interpretive tools and complex models of cultural interaction integral to humanities work may resolve varied and continuous issues. The authors suggest that a conceptual and physical shift toward interdisciplinarities within institutions (rather than interdisciplinary institutions, models, or methods) offers a realistic and flexible approach to transforming academia and education.

Drucker, Johanna. 2006. "Graphical Readings and the Visual Aesthetics of Textuality." *TEXT: An Interdisciplinary Annual of Textual Studies* 16: 267–76. http://www.jstor.org/stable/30227973.

Drucker discusses design aspects and graphic features that often go unnoticed in print, manuscript, electronic, and text formats. She states that the conception of design elements as autonomous entities is problematic, since it ignores the relational forms of expression in design systems. Drucker describes the space of the page as a system, or a quantum field, in which all graphical elements operate together in "a relational, dynamic, dialectically potential 'espace' constitutive of, not a pre-condition for, the graphical presentation of a text" (270–71). Defining the categories of graphic, pictorial, and textual space, Drucker performs a reading of a page from Boethius's *Consolatione* to demonstrate her proposed reading and interpretive approach to materiality in textual studies.

_____ 2011a. "Humanities Approaches to Graphical Display." *Digital Humanities Quarterly* 5 (1): n.p. http://www.digitalhumanities.org/dhq/vol/5/1/000091/000091.html.

Drucker proposes a usability and interaction design approach to data visualization in humanities fields. She draws attention to the fact that many digital visualization tools presuppose an observer-independent reality and an unquestionable representation. Counter to traditional humanities thinking, these tools do not acknowledge ambiguity, interpretation, or uncertainty. Drucker urges humanists to recognize all data as capta (which is actively taken rather than given). Furthermore, she advocates for forms of visual expression that display information as constructed by human motivation and perceived according to the interpretation of the viewer or reader. Her argument also opens up space for more 3D representations in data visualization, adding subjective experience to otherwise 2D expressions of time and space. Drucker stresses that such graphical approaches are imperative for humanities tenets to be applied and implemented in digital graphical expressions and interpretations.

*_____ 2011b. "Humanities Approaches to Interface Theory."
    Culture Machine 12: 1–20. http://www.culturemachine.net/index.
    php/cm/article/viewArticle/434.

In Drucker's humanities theory of interface, she argues that the interface
is the predominant site of cognition in digital spaces and requires cognizant, intellectual design. Drucker's theory is predicated on interface design
that considers the constitution of a subject, not the expected activities of
a user; on graphical reading practices and frame theory; on constructivist
approaches to cognition, and on integrating multiple modes of humanities
interpretation. She argues for a humanities approach to interface theory
that integrates different forms of reading and analysis in order to allow readers to recognize the relations of the dynamic space between environments
and cognitive events. Furthermore, while avoiding a descent into screen essentialism, Drucker insists that studying electronic reading practices must
be focalized through studying graphical user interfaces, as GUIs constitute
reading (and thus the reading subject, or "subject of interface" [3]).

*Guldi, Jo. 2013. "Reinventing the Academic Journal." In *Hacking the
    Academy: New Approaches to Scholarship and Teaching from Digi-
    tal Humanities*, edited by Daniel J. Cohen and Tom Scheinfeldt,
    19–24. Ann Arbor: University of Michigan Press. doi:10.3998/
    dh.12172434.0001.001.

Guldi calls for a rethinking of scholarly journal practices in light of the emergence and allowances of Web 2.0. She argues that journals can re-establish
themselves as forthright facilitators of knowledge creation if they adopt
notions of interoperability, curation, multimodal scholarship, open access,
networked expertise, and transparency regarding review and timelines. For
Guldi, the success of the academic journal depends on incorporating social
bookmarking tools and wiki formats. Journals should assume a progressive
attitude predicated on sharing and advancing knowledge instead of a limiting view based on exclusivity, profit, and intellectual authority.

*Hayles, N. Katherine. 2008. *Electronic Literature: New Horizons for the
    Literary*. Notre Dame, IN: University of Notre Dame Press.

Hayles surveys the field known as electronic literature. She suggests that
while electronic literature acknowledges the expectations formed by the
print medium, it also builds on and transforms them. In addition, electronic
literature is informed by other traditions in contemporary digital culture,
including computer games. In this way, electronic literature embodies a hybrid of various forms and traditions that may not usually fit together. Hayles
outlines a wide variety of electronic literature examples, and comments that

new approaches of analysis are required—in particular, the ability to "think digital" and to recognize the aspects of networked and programmable media that do not exist in print literature. In electronic literature, neither the body nor the machine should be given theoretical priority. Instead, Hayles argues for interconnections that "mediate between human and machine cognition" (x). She sees this intermediation as a more playful form of engaging with the complex mix of possibilities offered by contemporary electronic literature.

Huizinga, Johan. 1949. *Homo Ludens: A Study of the Play-element in Culture.* **London: Routledge and Kegan Paul.**
Huizinga offers a thorough study and analysis of forms of play. His definition and characteristics of play are widely cited among game scholars and other theorists, demonstrating the significance of his initiative to acknowledge the value of studying the meaning of play. Huizinga carefully outlines characteristics of play: play is a free activity; play steps outside of "real" life; play is different from ordinary life because it is restrained by locality and duration; play consists of rules and has order; and play includes no material interests or profit. While the definition of games and play remains a much-debated topic, Huizinga's categories offer an important starting point. One key term in contemporary game studies that has emerged from *Homo Ludens* is the concept of the magic circle: gameplay is isolated from "real" life through locality and duration—play starts and ends, and it is limited in terms of time and space. All play occurs within the realm of these playgrounds.

Jagoda, Patrick. 2014. "Gaming the Humanities." *Differences* **25 (1): 189–215.**
Jagoda claims that digital games require new critical approaches, as they have a wide impact on various twenty-first-century fields of study and practice. He reflects on games and gamification as a problematic tendency in the digital humanities, primarily through his project "Game Changer Chicago Design Lab." The Chicago Design Lab experiments with transmedia narrative, collaborative design, and engaging with digital humanities through new artistic forms. When discussing games and gamification, Jagoda acknowledges the cultural spread of games and the growth in participants and stakeholders, as well as the effect of gaming as a pedagogical technique. He elaborates on a case study of the reality game *The Source* (also included in this bibliography), which deals with varied social justice issues that include bullying, immigration, and health policy. Overall, Jagoda considers collaboration, social justice design research methods, and intersections between the humanities and the sciences, and concludes with the importance of gaming in the humanities as

a method of research and collaboration between humanists, artists, designers, technologists, scientists, and educators.

Jones, Steven E. 2011. "Performing the Social Text: Or, What I Learned from Playing Spore." *Common Knowledge* 17 (2): 283–91. doi:10.1215/0961754X-1187977.

Jones examines how texts and video games offer performative social system environments that allow for collaborative modelling toward knowledge development and acquisition. He sees video games as social objects that, similar to texts, attain their meaning only through engagement of the player or reader, where players take on a director/metaeditor role through content creation and content sharing. He describes the environment of the simulation game *Spore* "as a continually reedited universe of content-objects" (288). Jones goes on to compare game play in *Spore* to textual analysis, referring to Jerome McGann's development of *Ivanhoe* as an example, and considers the ways in which both areas allow for modelling to visualize interpretation and rewriting by players. He calls for a cyberinfrastructure for the humanities that allows for interpretive consequences within a social and a structural space. In this space, players/readers/textual analysts learn through complex, collaborative modelling, and knowledge is acquired through the process of manipulating representations. A textual editing environment based on this premise would remain purposefully unfixed, open, shared, and perpetually capable of manipulation.

_____ 2009. "Second Life, Video Games, and the Social Text." *PMLA* 124 (1): 264–72.

Jones considers the similarities between the metaverse space in games such as *Second Life* and the social text and Web 2.0 generally. He explains that in these game spaces tagged objects exist in relation to users (who may also be metatagged through technologies such as RFID chips), thus forming structures in which interactions unite users and objects. Jones argues that these social spaces do not exist apart from the "real world" of meaning making and production. In games such as *World of Warcraft, Second Life, Spore,* and *The Sims,* and in certain alternate-reality games (ARGs), collaborative construction is already taking place to create objects and information. Jones concludes that such video game spaces provide humanists with models of networked, metatagged, multi-dimensional environments.

*_____ 2013. *The Emergence of the Digital Humanities.* London and New York: Routledge.

Jones studies the emergence of digital humanities in response to changes in culture. He uses William Gibson's concept of the eversion of cyberspace (that is, the boundary crossing, flipping, and erasure between cyberspace and non-cyberspace) as a way to describe the cultural change that has led to the current incarnation of digital humanities. Jones frames the emergence of digital humanities as a blending of textual studies and game studies. He provides readings of popular games such as *Fez* and *Spore*, as well as a number of indie games, to analyze the relation between digital humanities and game studies. Jones concludes with an overview of practices (such as desktop fabrication) that are relevant to both gaming and digital humanities. For a snapshot of Jones's stated views on scholarly communication, please see the annotation of the "Publication" chapter from *The Emergence of the Digital Humanities* elsewhere in this bibliography collection.

*Latour, Bruno. 2009. "A Cautious Prometheus? A Few Steps Towards a Philosophy of Design (with Special Attention to Peter Sloterdijk)." In *Networks of Design: Proceedings of the 2008 Annual International Conference of the Design History Society*, edited by Fiona Hackne, Jonathan Glynne, and Viv Minto, 2–10. Boca Raton, FL: Universal Publishers.

Latour meditates on the form and function of the term "design," and proposes a more comprehensive vision for the practice. He suggests that design practitioners focus more fully on drawing together, modelling, or simulating complexity—more inclusive visions that incorporate contradiction and controversy. Latour argues that we are living in an age of design (or redesign) instead of a revolutionary modernist era of breaking with the past and making everything new. Increasingly, design encapsulates various other acts, from arrangement to definition, from projecting to coding. Consequently, the possibilities and instances for design grow exponentially. For Latour, the concept of an age of design predicates an advantageous condition defined by humility and modesty (because it is not foundational or construction-based); a necessary attentiveness to details and skillfulness; a focus on purposeful development (or on the meaning of what is being designed); thoughtful remediation; and an ethical dimension (exemplified through the good design versus bad design binary).

*Losh, Elizabeth. 2012. "Hacktivism and the Humanities: Programming Protest in the Era of the Digital University." In *Debates in the Digital Humanities*, edited by Matthew K. Gold, 161–86. Minneapolis: University of Minnesota Press. http://dhdebates.gc.cuny.edu/debates/text/32.

Losh scans the instantiations of, and relations between, hacktivism and the humanities. She contends, along with scholar Alan Liu, that through an

increased self-awareness the digital humanities can actually effect real political, social, public, and institutional change. Losh examines the hacking
rhetoric and actions of scholar Cathy Davidson (via the HASTAC collaboratory); the Radical Software Group and its director Alexander Galloway; and
the Critical Art Ensemble, with a focus on CAE member and professor Ricardo
Dominguez. Losh concludes by acknowledging criticism of the digital humanities, and suggests a solution: digital humanists should engage in more
public, political collaborations and conversations.

McGann, Jerome. 2001. *Radiant Textuality: Literature after the World Wide
    Web.* New York: Palgrave.
McGann's compilation of essays from 1993 to 2000 shows the development
of his work in the digital edition, literary studies and interpretation, and
digital scholarly work. He comes to regard critical gaming structures as
environments that allow for new approaches to these areas of study. The
essays move through McGann's understanding of the potential of digital
technologies as "thinking machines" that can go beyond the material limitations of the book. He describes scholarly work, editions, and translations as
performative deformation that manipulates text and supplies a perceptual
presentation for the reader. McGann explores the opportunity to leverage
the digital ecosystem and enable interplay between multiple fields by using
markup and databases to make "N-dimensional space" accessible. The final
chapter reveals how the digital game *Ivanhoe* offers such an environment. In
Ivanhoe, a digital role-playing game, a literary work is read and interpreted
in a framework that combines primary and secondary texts, scholarship,
and the players' interpretations and commentaries in the same area, thus
encouraging new forms of critical reflection. McGann calls this a "quantum
field," where textual objects and reading subjects operate within the same
space, and which allows for algorithmic and rhetorical performative activity
within rather than outside of the object of attention.

*Ramsay, Stephen, and Geoffrey Rockwell. 2012. "Developing
    Things: Notes Toward an Epistemology of Building in the Digital
    Humanities." In *Debates in the Digital Humanities*, edited by Matthew
    K. Gold, 75–84. Minneapolis: University of Minnesota Press. http://
    dhdebates.gc.cuny.edu/debates/part/3.
Ramsay and Rockwell take up the "your database/prototype is an argument"
conversation (notably championed by Lev Manovich and Willard McCarty).
They assert that taking building as seriously as scholarly work could productively dismantle or realign the focus of the humanities from its predominantly textual bent. Ramsay and Rockwell advocate for installing the user,

reader, or subject at the level of building. Through this socially minded conceptual and physical shift, some of the abstractions and black boxing that render digital humanities tools theoretically insufficient could be avoided or amended.

Ryan, Marie-Laure. 1994. "Immersion *vs*. Interactivity: Virtual Reality and Literary Theory." *SubStance* 28 (2): 110–37. doi:10.1353/ sub.1999.0015.

Ryan examines the theoretical implications of virtual reality (VR) in relation to literary theory. She notes the similarities between literary devices commonly used to create a sense of reader participation in a fictional world, and the immersion and interaction devices used in VR to effect what Ryan calls "telepresence." She identifies immersion (the realistic representation) and interaction or interactivity (the ability to not only navigate but modify) as the two key features that create experiences of reality. Ryan considers VR a semiotic phenomenon, and states that the VR effect is the "denial of the role of signs," thus allowing for an unmediated environment by working toward the appearance of a transparent medium. She concludes that textual environments are limited in their ability to develop experiences of reality in the way VR does, because their tools of interactivity remain signs instead of physical, unmediated interactivity passing through the body.

*Vetch, Paul. 2010. "From Edition to Experience: Feeling the Way towards User-Focused Interfaces." In *Electronic Publishing: Politics and Pragmatics*, edited by Gabriel Egan, 171–84. New Technologies in Medieval and Renaissance Studies 2. Tempe, AZ: Iter Inc., in collaboration with the Arizona Center for Medieval and Renaissance Studies.

Vetch explores the nuances of a user-focused approach to scholarly digital projects, arguing that the prevalence of Web 2.0 practices and standards requires scholars to rethink the design of scholarly digital editions. For Vetch, editorial teams need to shift their focus to questions concerning the user. For instance, how will users customize their experience of the digital edition? What new forms of knowledge can develop from these interactions? Moreover, how can rethinking interface design of scholarly digital editions promote more user engagement and interest? Vetch concludes that a user-focused approach is necessary for the success of scholarly publication in a constantly shifting digital world.

2. Game-Design-Inspired Learning Initiatives

Carson, Stephen, and Jan Philipp Schmidt. 2012. "The Massive Open Online Professor." *Academic Matters: The Journal of Higher Education*, n.p. http://www.academicmatters.ca/2012/05/the-massive-open-online-professor/.

Carson and Schmidt offer an overview of the current state and possible effects of Massive Open Online Courses (MOOCs). MOOCs were initiated by academic institutions such as Stanford and MIT, and they offer free, online courses that hundreds of thousands of users can enroll in at minimal additional cost to the institution. The authors describe the characteristics of MOOCs as consisting of open content, peer-to-peer interactions, automated assessment and grading, and alternative recognition or credential systems. Gamification, and specifically the use of badges, has been an approach led by the Mozilla Foundation, the MacArthur Foundation, and Peer 2 Peer University to develop a new way of acknowledging learning achievements. Carson and Schmidt speculate about the lasting changes MOOCs may bring about, such as the possibility of long-term engagement in learning (beyond the completion of university courses and degrees).

Danforth, Liz. 2011. "Gamification and Libraries." *Library Journal* 136 (3): 84–85.

Danforth defines gamification as the application of gameplay mechanics in non-game settings. She contextualizes gamification as a method often used in marketing tactics in a type of rewards-based incentive program. Danforth acknowledges that gamification can be beneficial if it is engaging and encourages creative thinking. She points out its employment in educational settings and sees gamification's potential use in enhancing library skills and intellectual endeavours.

Dickey, Michele D. 2007. "Game Design and Learning: A Conjectural Analysis of How Massively Multiple Online Role-Playing Games (MMORPGs) Foster Intrinsic Motivation." *Educational Technology Research and Development* 55 (3): 253–73.

Dickey investigates how massively multiple online role-playing games (MMORPGs) may offer structural models for the design of interactive learning environments. She focuses on the aspects that support intrinsic motivation in MMORPGs: character design, narrative, and player motivation, as well as how scaffolding for problem solving and narrative structure encourage learning. Dickey conducts a thorough literature review and recognizes that MMORPGs are structured as collaborative, strategy-driven, multimodal, interactive environments. These attributes tie in to the objectives of interactive learning environments that seek to generate collaboration and critical thinking.

Gibson, David, Clark Aldrich, and Marc Prensky, eds. 2007. *Games and Simulations in Online Learning: Research and Development Frameworks.* Hershey, PA: Information Science Publishing.
Gibson, Aldrich, and Prensky's compilation of essays provides a thorough overview of the opportunities that games and simulations offer in the design of online learning environments. The collection covers an array of areas, such as innovative design models, game-based assessment, learning and instruction in networked virtual worlds, and Massively Multiplayer Online Games (MMOGs). Essays also include discussion on the use of simulation for discovery learning, guidelines for the development of prototypes and applications that include game and simulation approaches, and the analytics capabilities offered by game and simulation approaches in online education. The editors incorporate a wide scope of content for scholars and instructors who work in different fields and education stages.

Jensen, Matthew. 2012. "Engaging the Learner." *Training and Development* 66 (1): 40.
Jensen outlines approaches, practices, and risks in using gamification for learning environments. He notes that successful gamification must elicit meaningful engagement by prioritizing the player experience, including by making the experience personally relevant and gearing it toward the target audience. He also highlights the power of narrative. Player-centred games in a successful gamification environment share the common characteristics of being responsive, collaborative, ritualistic, incremental, convenient, and rewarding. Thus, gamification should be approached by thinking like game designers, rather than simply implementing decontextualized mechanisms.

Jing, Tee Wee, Yue Wong Seng, and Raja Kumar Murugesan. 2015. "Learning Outcome Enhancement via Serious Game: Implementing Game-Based Learning Framework in Blended Learning Environment." *5th International Conference on IT Convergence and Security (ICITCS)*, 24–27 August. doi:10.1109/ICITCS.2015.7292992.
Jing, Seng, and Murugesan discuss how serious games have emerged as a new learning strategy because of the need for engaging and interactive educational practices. They integrate pedagogy and game-design strategy and implement the game-based learning (GBL) framework in blended learning environments. The authors rely on instructional design theories and

educational game-design models in their study. Their framework can be referred to for the identification and refinement of key elements in serious games. Jing, Seng, and Murugesan conclude that a proper implementation of the GBL framework in serious games will increase students' interest and enhance their academic outcomes.

Kapp, Karl M. 2012. *The Gamification of Learning and Instruction: Game-Based Methods and Strategies for Training and Education.* San Francisco: Pfeiffer.

Kapp offers a practical guide for readers who want to implement gamification in learning environments. He provides definitions and examples of gamification, surveys individual elements and aspects of gamification and reviews them in detail, discusses the different levels of effectiveness of gamification for instructional purposes, and offers practical advice to planning the development of a gamified learning environment. Kapp is critical of common implementations of gamification (e.g., merely placing badges into a tool, trivializing learning, or only considering basic game mechanics rather than actual game-design practices). His detailed analysis and overview of gamification methods to improve learning environments provides educators and scholars with a thorough resource on the topic.

Mysirlaki, Sofia, and Fotini Paraskeva. 2012. "Leadership in MMOGs: A Field of Research on Virtual Teams." *Electronic Journal of E-Learning* 10 (2): 223–34.

Mysirlaki and Paraskeva develop a theoretical framework for the analysis of leadership and social interactions in Massively Multiplayer Online Games (MMOGs) and Massively Multiplayer Online Role-Playing Games (MMORPGs). Recognizing these environments as self-organized, complex systems, the authors consider how the social structures of MMOGs and MMORPGs may offer insights for the design of collaborative virtual environments. The authors focus specifically on leadership skills and how a sense of community is related to player motivation.

Squire, Kurt. 2008. "Open-Ended Video Games: A Model For Developing Learning for the Interactive Age." In *The Ecology of Games: Connecting Youth, Games, And Learning,* edited by Katie Salen, 167–98. Cambridge, MA: MIT Press.

Squire reviews different types of video games, including targeted games, epistemic games, and augmented reality role-playing games. He focuses his analysis on open-ended simulation games, or sandbox games, as theoretical models for video game-based learning environments. Taking *Civilization* and

Grand Theft Auto: San Andreas as examples, he looks at identity, competitive spaces, and experiences within those spaces, before moving on to consider more education-related insights. Squire considers how games are designed as communities for learning, the forms of engagement in open-ended games in school settings, interpretations of history through games, games as learning systems, and participatory education. Based on the insights gained from this review, Squire concludes that sandbox game approaches offer educators new models and forms to enable student participation and learning.

Van Staalduinen, Jan-Paul, and Sara de Freitas. 2011. "A Game-Based Learning Framework: Linking Game Design and Learning Outcomes." In *Learning to Play: Exploring the Future of Education with Video Games*, edited by Myint Swe Khine, 29–54. New York: Peter Lang.
Van Staalduinen and de Freitas suggest that there is a need for new design methods that have an effect in both academic and formal contexts. The authors propose a hybrid framework that can be used in the design of new games, as well as in the analysis of existing game designs; it can also serve as a guide for designing new, more effective serious games. They lay out the limitations of the field of learning from video games, then acknowledge what they consider to be good instructional games. Van Staaduinen and de Freitas conclude that the learner has an important role to play in the dialogue between game design and learning, yet there needs to be greater assessment of the efficiency of learning-driven game-design strategy.

3. Game-Design Models in the Context of Social Knowledge Creation Tools

Blizzard Entertainment, Inc. 2005. *World of Warcraft* (WoW) [video game]. https://worldofwarcraft.com/en-us/.
World of Warcraft (WoW) is the world's most subscribed-to Massively Multiplayer Online Role-Playing Game (MMORPG). Set in the universe of Warcraft, players create avatars based on different races and characters. Gameplay can consist of quests assigned by non-player characters (NPCs), setting up player-versus-environment (PvE) gameplay, or players can engage in player-versus-player combat (PvP). While WoW players can play solely as individuals, the formation of guilds and subsequent strategic play is common.

CCP Games. 2003. *EVE Online* [video game]. http://www.eveonline.com/.
EVE Online is a multiplayer MMORPG that takes place in a science fiction space setting. Players can assume or create one or multiple characters to navigate a galaxy set 21,000 years in the future. The galaxy consists of more than 7,500 star systems that players can navigate in space ships, accessing different star

systems by means of stargates. Characters can take on different races and societies, and they can engage in different professions and activities, such as mining, trading, manufacturing, piracy, and combat. *EVE Online* consists of a large community of subscribers, which surpassed 500,000 in 2013.

Chang, Edmond. 2012. "Video+Game+Other+Media: Video Games and Re-
    mediation" [blog post]. *Critical Gaming Project.* https://depts.wash-
    ington.edu/critgame/wordpress/2012/01/videogameothermedia-
    video-games-and-remediation/.

Chang considers video games within media culture and the adaptation of games for other purposes in the context of remediation. Referring to his work with Sarah Kremen-Hicks, Chang questions whether we can imagine new media only in the frame of old media and in existing structures of information. He argues that innovation in a medium can only be based on prior innovation of technology. Within this framework, innovation may not necessarily create better products, only more products, which indicates the teleological refinement that takes place and recognizes the "effect of new forms on existing ones" (n.p.).

Center for Interdisciplinary Inquiry and Innovation in Sexual and
    Reproductive Health (Ci3), University of Chicago. 2013. *The Source*
    [video game]. https://ci3.uchicago.edu/portfolio/the-source/.

The Source was an alternate reality game created as part of a STEM (Science, Technology, Engineering, and Math) high school curriculum; it was played by some 140 youth across the Chicago area from July 8 to August 16, 2013, during that year's Chicago Summer of Learning. The game consisted of a series of webisodes showing Adia, a 17-year-old African American girl, speaking through her webcam to the players. Players split up into teams to solve problems and to help Adia understand a letter she had received. Through this process, the youth playing the game engaged in investigation and code-breaking, solved STEM-based puzzles, and participated in media production.

Crowley, Dennis, and Naveen Selvadurai. 2009. *Foursquare.* New York:
    Foursquare. https://foursquare.com.

Foursquare is a location-based social networking application primarily developed for mobile use. The main activity consists of users checking into different locations and tagging either the venue or the activity. Foursquare is built as a gamified structural mechanism that is often used as a model for gamification. Every check-in helps the user gain points, and certain tags or specific locations can earn the user badges. Users can become "mayors" of certain locations if they check in more than any other user over a certain time span.

Cyber Creations Inc. 2002. MMORPG.*com* [video game website]. http://www.mmorpg.com/.
Headed by Craig McGregor, *MMORPG.com* is the website for enthusiasts of Massively Multiplayer Online Role Playing Games. MMORPGs are usually quest-oriented, and often reward players for working together. Almost all MMORPGS include elements of currency and trade, which often encourages players to exchange items with or buy items from each other in a virtual free market. Many of these games will encourage cooperative play through quests or missions that require more than one player to complete. MMORPGs may also have built-in multiplayer minigames or player competitions. Notable titles: *World of Warcraft, The Lord of the Rings Online: Shadows of Angmar, The Elder Scrolls Online,* and *RuneScape.*

De Carvalho, Carlos Rosemberg Maia, and Elizabeth S. Furtado. 2012. "Wikimarks: An Approach Proposition for Generating Collaborative, Structured Content from Social Networking Sharing on the Web." In *Proceedings of the 11th Brazilian Symposium on Human Factors in Computing Systems (IHC '12),* 95–98. Porto Alegre, Brazil: Brazilian Computer Society.
De Carvalho and Furtado argue for what they call a Wikimarks approach in order to encourage organized, sustainable, social content creation. Based on this approach, users share online content that flows into a content repository and is subsequently categorized in a taxonomy system by the users. User participation is fostered through social interaction and extrinsic motivation. In order to motivate participation in the classification of content, the authors recommend gamification methods.

De Paoli, Stefano, Nicolò De Uffici, and Vincenzo D'Andrea. 2012. "Designing Badges for a Civic Media Platform: Reputation and Named Levels." In *Proceedings of the 26th Annual BCS Interaction Specialist Group Conference on People and Computers (BCS-HCI '12).* Swinton, UK: British Computer Society, 59–68.
De Paoli, De Uffici, and D'Andrea outline a design experience for badges in Civic Media Platforms (CMPs) based on insights gained from a CMP design model called "timu," which aims to offer a framework for a participative, bottom-up information ecosystem. While they acknowledge critiques of gamification, they argue that badges offer a way to formalize skills and reputation. The authors review various strengths and opportunities that badges

bring to civic and educational platforms: they can represent a number of different things (e.g., community membership, competence, experience, reputation); they support transferability of skills, reputation, or achievements; they trigger motivation; and they build a sense of community among participants.

Drucker, Johanna. 2003. "Designing Ivanhoe." *TEXT Technology* 2: 19–41.
Drucker charts the interface design approach that was used in the development of the *Ivanhoe* project that she worked on with scholar Jerome McGann. The objective was to challenge usual design practices and their assumptions about clarity and communication. Instead of designing *Ivanhoe* based on the structuralist premise that visual presence and graphical form are self-evident, Drucker used a theory-driven approach that allows for the interface to be conceived of as dialogic and networked, generative and procedural, emergent, relational, iterative, dialectical, and transformative. *Ivanhoe* is designed so that critical awareness is not only a part of the game (through the textual studies perspective), but the interface itself is based on critical awareness and theoretical insights.

Drucker, Johanna, and Jerome McGann. 2000. *Ivanhoe*. SpecLab (University of Virginia). http://www.ivanhoegame.org/?page_id=21.
Ivanhoe is an online game environment for multiple readers to collaboratively read, interpret, and reflect on a literary text. Similar to other role-playing game (RPG) environments, players take on alternate identities to perform their reading and interactions with each other. This structure encourages players to be aware of the ways in which acts of interpretation are formed, encouraging reflection on the meaning of such acts. Thus, the game enables collaborative interpretation of the selected text as well as critical reflection of the interpretive process itself. The gamespace, or bookspace, consists not only of the primary literary text that the game is structured around, but combines multiple primary and secondary texts, player contributions, and computer-generated processes in the same sphere.

Galloway, Alexander R., Carolyn Kane, Adam Parrish, Daniel Perlin, DJ /rupture and Matt Shadetek, Mushon Zer-Aviv, and the RSG Collective. *Kriegspiel* [video game]. http://r-s-g.org/kriegspiel/index.php.
Galloway and the RSG Collective of programmers and artists designed *Kriegspiel* ("war game" in German) based on Guy Debord's board game of the same name. Debord first produced a limited edition of the game in 1977. He developed a full rule book, a mass production model made of cardboard and wood tiles, and a book that he co-published with his wife, Alice Becker-Ho, in

1987. *Kriegspiel* is a variant of chess played by two opponents—each of whom controls an army—on a 500-square board arranged in 20 rows of 25 squares each. The digital game is an attempt to situate Debord's game in a contemporary landscape.

Insemtives. 2009–13. *INSEMTIVES.* **http://insemtives.eu/.**
This suite of tools focuses on the creation of semantic content via incentive-based gaming environments. INSEMTIVES aims to bridge the gap between machine-readable computational data and the necessary limitations of automating semantic content creation tasks. By providing incentives, INSEMTIVES attempts to inspire individuals to manually create, extend, or revise semantic content. This tool is geared toward social knowledge creation through user-generated content and participation.

Jakobsson, Mikael. 2011. "The Achievement Machine: Understanding Xbox 360 Achievements in Gaming Practices." *Game Studies: The International Journal of Computer Game Research* **11 (1): n.p. http:// gamestudies.org/1101/articles/jakobsson.**
Jakobsson scans the achievements system in Xbox 360 games. In this console gaming environment, multiple individual games are combined into a total score or achievement level that is visible to other players, similar to the structure of massively multiplayer online (MMO) game environments. The achievements system provides extrinsic rewards that can be seen by others and thus function as external motivators. Comparing MMO game environments and console gaming, Jakobsson notes that both have similar properties, such as persistence, coveillance, and open-endedness. Jakobsson concludes that, although the achievements system in Xbox games follows rewards system approaches, it functions like an MMO game that all Xbox Live members participate in.

Kopas, Merritt. 2012. *Lim* **[video game]. New York: Games for Change. http://www.gamesforchange.org/play/lim/.**
Kopas's game *Lim* requires the player to move a square through a structure of other squares (using the arrow keys) and to take on the colour of other squares in order to fit in and avoid attack. Built in Construct 2, a DIY game-making platform, *Lim* offers a superb example of the ways in which game mechanics can make arguments. While highly abstract, the game clearly communicates certain feelings, such as those of distress and not fitting in, that are important to the topic of liminality.

Maxis. 2008. *Spore* **[video game]. Redwood City, CA: Electronic Arts Inc.**
    http://www.spore.com/.
Spore is a multi-genre, single-player god game wherein the player develops
a species and aims to achieve certain objectives in different stages of its development. The way each stage is played determines new characteristics that
the species obtains for the following level. *Spore* consists of several genres,
including action, strategy, and role playing game (RPG). The species that
players create can be loaded to Sporepedia online, allowing other players to
download them.

Maxis and The Sims Studio. 2000/2006–. *The Sims* **[video game]. Redwood**
    City, CA: Electronic Arts Inc. http://www.thesims.com/en-us.
The Sims is a best-selling strategic life simulation video game that consists of
a main series and a variety of spinoffs. It is structured as a sandbox game in
which players create people called "Sims." The gameplay consists of helping
these "Sims" live in their houses, engage in daily activities, and satisfy their
desires.

McGann, Jerome. 2005. "Like Leaving The Nile. IVANHOE, A User's
    Manual." *Literature Compass* **2: 1–27. http://www2.iath.virginia.**
    edu/jjm2f/old/compass.pdf.
In this user manual for the online literary analysis game *Ivanhoe*, McGann explains why he considers it imperative that humanities activities such as text
analysis and interpretation move into and embrace the digital realm. While
recognizing that humanities and social sciences material must be treated as
information at the computational level, he argues that such materials must
also be treated as knowledge at the "level of perception and thought—at
the level of their human uses" (4). In *Ivanhoe*, an online gamespace, multiple readers can explore and interpret a text in a manner that visualizes
the interpretations and shows interrelations between the players, moves,
and documents. *Ivanhoe* thus allows for interpretation to take place on two
levels: through interpretation of the documents being studied, and through
interpretation of the critical thought of the players participating. McGann
explains the functions and interactions of the game by walking through a
textual mockup of an actual gameplay as an example.

Meier, Sid. 1991. *Civilization* **[video game]. Hunt Valley, MD: MicroProse.**
    http://www.civilization.com.
Civilization is a turn-based strategy game in which players construct, control, develop, and manage an empire. The player rules the civilization, builds
cities and expands the empire, and at times has to engage in warfare and

protect the empire. The culture, technology, and intellectual state of the civilization develops as the empire evolves. There have been six iterations of *Civilization*, including the original, as well as numerous expansion packs, spinoffs, and even board and card game versions.

Mojang and Microsoft Studios. 2011. *Minecraft* [video game]. Stockholm: Mojang; Redmond, WA: Microsoft Studios. https://minecraft.net.
Originally created by Swedish designer Markus "Notch" Persson, *Minecraft* is an open-world or sandbox game that allows for players to engage in activities outside of specific goals. The main activity in the game is to build constructions within a grid system using blocks that consist of a variety of materials. Most often, players play in the first person, but *Minecraft* also allows for third person gameplay. The game contains an optional achievement system, and players can choose to play in a survival mode or a creative mode, thus enabling different types of activities.

Multiplayer Online Battle Arena (MOBA) games. 2009–.
Since coming to prominence in 2009, the Multiplayer Online Battle Arena (MOBA) style of game has grown in popularity due largely to the success of two titles: *League of Legends* and *Smite*. In MOBA games, whole team cooperation and stratification is paramount to success. Every player has a specific role in this style of game, and if one person is not performing well it can cost the whole team points, or result in the team losing the game. Often, the goal of a MOBA game centres around destroying one or more enemy structures while fighting off Artificial Intelligence-controlled units as well as the other team. Players usually play on a map that is split into lanes and jungle areas. They will then take designated positions and roles across the map, and cooperatively strategize on how to defeat the enemy. Notable titles: *League of Legends, Smite,* and *Defence of the Ancients.*

Polytron Corporation. 2012. *Fez* [video game]. http://fezgame.com/.
Fez is an indie puzzle and platform game developed by Montreal-based Polytron for Xbox Live Arcade. The game is unique in that it is a 3D world played from a 2D perspective. Gomez, the player character, starts out in a 2D world, but he receives a hat that allows him to enter the third dimension. Thus the player can rotate 90 degrees across four sides of the world to move through it. The goal of the game consists of collecting 32 cubes to reconstruct the hexahedron that existed in Gomez's world at the beginning of the game. In this pursuit, the player moves through the world, finds secrets, and solves puzzles; however, Gomez does not fight enemies, and, although death can occur, there is no penalty for it.

Real-Time Strategy Games (RTSGs). 1982–.
Making its debut in 1982, the Real-Time Strategy Game (RTSG) is a single-player, cooperative, or online strategy game played, as its name suggests, in real time. Often, RTSG players balance a country's or group's economy, and possibly their military as well. A prime RTSG example and notable title is *Total War: Shogun 2*, which is a detailed, turn-based strategy game as well as a real-time battle strategy game. In the Co-op campaign within *Shogun 2*, two players are tasked with taking over 24 provinces in Japan. Together, the players manage their clan's food, money, reputation, population, and military. When the players engage other groups in battle they do so via a third person view of their troops, and the two players must cooperatively battle the opposition's army in order to win.

Rockwell, Geoffrey. 2003. "Serious Play at Hand: Is Gaming Serious Research in the Humanities?" *TEXT Technology* 2: 89–99.
Rockwell examines the role of games in academic research within the humanities. Referring to the ideas of theorists Ludwig Wittgenstein, Johan Huizinga, Hans-Georg Gadamer, and others, Rockwell conducts an investigation of the game *Ivanhoe* (a game environment for collaborative interpretation of literary texts) to show how the humanities can combine gaming and research. He depicts *Ivanhoe* as a model game environment that enables "what criticism should and could be in the context of learning and collaborative research" (93), while bringing playfulness into humanities activities.

Rockstar Games, Inc. 1997. *Grand Theft Auto* (GTA) [video game]. http://www.rockstargames.com/grandtheftauto/.
The *Grand Theft Auto* (GTA) series is an open-world action-adventure driving game. Players take on characters who usually try to rise in the ranks of organized crime. Structured as a sandbox game, GTA is set in urban environments with fictional names, although they are based on American cities and states. The game action is primarily organized around vehicles, drivers, pedestrians, and traffic signals. However, gameplay goes far beyond driving, and player characters can choose which missions they complete and how they interact with other characters.

Stack Exchange Network. 2013. *Stack Overflow*. http://stackoverflow.com.
Stack Overflow is a free programming question and answer site that allows users to build their reputation in order to gain more access and privileges. The site aims to offer an environment that allows programmers to ask relevant questions and receive helpful answers while discouraging irrelevant content. Structured as a user-built and -run environment, the curation and development

of relevant content is encouraged through gamification methods. Within the Q&A framework, users can vote up each other's contributions, and the best answers are displayed at the top of the list of responses. As a user's questions, answers, and edits are voted up, that person's reputation score increases. The higher the reputation score, the higher the user's access privileges. Users can also earn badges for certain achievements and forms of participation.

Zynga Inc. 2009. *FarmVille* **[video game]. Facebook and HTML 5. http:// company.zynga.com/games/farmville.**
FarmVille is a social network game that leverages the Facebook environment. Gameplay consists of the management of a farm that players maintain by plowing land, raising livestock, and planting, growing, and harvesting crops. Each player has an avatar and can interact with friends through Facebook. Players earn farm coins through certain actions or by obtaining enough experience points to move up levels, or farm points can be purchased for real money. Players are encouraged to interact with friends by visiting each other's farms or joining efforts by forming co-ops. Ian Bogost's game *Cow Clicker*, launched in 2010, satirizes *FarmVille* and similar games.

4. Defining Gamification and Other Game-Design Models

Bogost, Ian. 2011. "Persuasive Games: Exploitationware" [blog post]. *Gamasutra.* **http://www.gamasutra.com/view/feature/6366/per- suasive_games_exploitationware.php.**
Bogost asserts that the power of gamification lies in the term's rhetorical effect, which diminishes how difficult games actually are and simplifies the field of gaming to make it applicable in multiple contexts. He states that gamification, as it currently appears in corporate and marketing platforms, should be replaced with the term "exploitationware," since it substitutes real incentives with fictional ones, thus creating exploitative relationships between company and consumer. In his pursuit to rid the industry of exploitative gamification, Bogost invokes the term "games-as-systems" to supersede gamification with alternatives that do "real, meaningful things with games" (n.p.).

Deterding, Sebastian, Dan Dixon, Rilla Khalad, and Lennart E. Nacke. 2011. "From Game Design Elements to Gamefulness: Defining 'Gamification.'" In *Proceedings of the 15th International Academic MindTrek Conference: Envisioning Future Media Environments (MindTrek '11)*, **9–15. New York: ACM. doi:10.1145/2181037.2181040.**
Deterding, Dixon, Khalad, and Nacke investigate gamification methods in order to define gamification and contrast it to other concepts such as pervasive

games, alternate reality games, and serious games. The authors outline the industry origins and precursors of gamification to indicate how contentious the term is. They define gamification as "the use of game design elements in non-game contexts" (9), because this definition focuses on games, not play; indicates that it consists of elements of games, rather than being structured as full games, as serious games would be; constricts gamification to game-design elements, rather than game-based technologies or practices; and contextualizes gamification outside of games for pure entertainment. They suggest that "gameful design" may be a better term to use in place of "gamification" within academic discourses.

Douma, Michael. 2011. "What is Gamification?" [blog post]. *Idea.* http:// www.idea.org/blog/2011/10/20/what-is-gamification/.
Douma defines gamification as "adapting game mechanics into a non-game setting — such as building online communities, education and outreach, marketing, or building educational apps" (n.p.). While differentiating between gamification, serious games, and playful interaction, Douma does allow for some leeway as to what defines gamification. He outlines numerous ideas and approaches for gamification, such as levels, cascading information theory, community collaboration, loss aversion, quests/challenges, and infinite gameplay. Badges, trophies, and points are discussed in the most detail. He notes that badges offer psychological functions such as setting goals, instruction, reputation, status and affirmation, and group identification, but that in addition to serving as external motivators, badges also need to be a part of a narrative and offer personalized, goal-oriented engagement.

Graham, Adam. 2012. "Gamification: Where's the Fun in That?" [blog post]. *Campaign* 43: 47. http://www.campaignlive.co.uk/article/1156994/gamification-wheres-fun-that.
Graham defines gamification as "the use of game thinking and game mechanics to enhance non-game contexts" (n.p.). For Graham, the skillful use of game elements makes it possible to increase engagement across varied applications. While he notes that it is possible to gamify anything, the majority of gamification examples simply follow a formulaic pattern set by the Foursquare model, which uses points, badges, leaderboards, and prizes as incentives for participation. Instead of following this process, Graham urges practitioners to consider the extensive array of game-design approaches available, and to determine which ones would be the most successful in inciting player flow based on the target audience's triggers and motivators.

Groh, Fabian. 2012. "Gamification: State of the Art Definition and Utilization." In *Proceedings of the 4th Seminar on Research Trends in Media Informatics (RTMI '12)*, edited by Naim Asaj, et al. Ulm, Germany: Institute of Media Informatics, Ulm University, 39–46. http://hubscher.org/roland/courses/hf765/readings/Groh_2012. pdf.

Groh reviews the definition of gamification developed by Deterding et al. (2011) and analyzes the opportunities and problems gamification offers in the context of self-determination theory. He points out the differences between game (ludus) and play (paidia), differentiates gamification from serious games (that is, full-fledged games for non-entertainment purposes rather than game elements), and notes how such game-design elements can be used to enhance other applications. Groh presents the ways in which the values of relatedness, competence, and autonomy inherent in self-determination theory are also key components for gamification to be effective.

Jagoda, Patrick. 2013. "Gamification and Other Forms of Play." *Boundary 2* 40 (2): 113–44. doi:10.1215/01903659–2151821.

Jagoda discusses the ubiquity of games in different digital contexts, and explores gamification in particular. Defining gamification as "the use of game mechanics in traditionally nongame activities" (114), Jagoda sees gamification as an approach that uses game mechanics and objectives to function as an interface between work, leisure, thought patterns, affects, and social relations common in the current overdeveloped world and "the real." This gamified world, Jagoda argues, differs from a society oriented around the production of what Guy Debord called "spectacles." Rather than relying on unidirectional representations, the gamified world is structured in a bidirectional, many-to-many format that encourages engagement through customization and user-generated content. While Jagoda acknowledges that gamification may perpetuate a capitalist hierarchy, he also notes that game-based approaches can function to resist those socioeconomic structures. He analyses three games that problematize gamification: *SPENT* (2011), *Third World Farmer* (2006), and *Thresholdland* (2010). These games, rather than perpetuating a false sense of triumph and winning, draw attention to the failure that the majority of people experience in contemporary capitalism, thus functioning as critiques not only of the capitalist system but also of gamification. Jagoda demonstrates that although games and gamification often perpetuate dominant socioeconomic hierarchies and exploitation, game-based approaches can also function as forms of resistance.

Ritterfeld, Ute, Michael Cody, and Peter Vorderer, eds. 2009. *Serious Games: Mechanisms and Effects*. New York and London: Routledge.
Ritterfeld, Cody, and Vorderer explore how games can encourage learning in the real world. The editors define serious games as "any form of interactive computer-based game software for one or multiple players to be used on any platform and that has been developed with the intention to be more than entertainment" (6). Organized into four sections, the book's chapters explore the psychological mechanisms of serious games and how they facilitate learning, development, and change in a variety of areas, including health care, human rights, education, research, and immigration.

Rose, Frank. 2011. *The Art of Immersion: How the Digital Generation is Remaking Hollywood, Madison Avenue, and the Way We Tell Stories*. New York: W.W. Norton.
Rose explores how the Internet changes storytelling. He argues that while stories in other media also appear in patterns that we make meaning out of, the Internet communicates narratives in a unique way, changing how we communicate, create, consume, and engage with content. Rather than communicating stories as sequential narratives, the Internet allows for stories to be communicated in a nonlinear, participatory, game-like, and immersive way. This allows for deeper engagement with stories, especially when distinctions between author and audience, story and game, entertainment and marketing, and fiction and reality become increasingly blurred.

Werbach, Kevin. 2014. "(Re)Defining Gamification: A Process Approach." In *Proceedings of the 9th International Conference on Persuasive Technology (PERSUASIVE 2014), Lecture Notes in Computer Science* 8462: 266–72. doi:10.1007/978-3-319-07127-5_23.
Werbach defines gamification as a process of making activities game-like, focusing on the components of a game and the experience of gamefulness. He stresses that gamification is a process, so as to avoid the need of pointing to where the designed system crosses into becoming gamification. He calls upon the literature on persuasive design, since gamification influences behaviour. Werbach concludes that how one defines gamification will affect its coherence and shape a debate of legitimacy, suggesting that his definition captures the essential aspects of the practice, while providing direction for its future.

Zichermann, Gabe, and Christopher Cunningham. 2011. *Gamification By Design: Implementing Game Mechanics in Web and Mobile Apps.* **Sebastopol, CA: O'Reilly Media.**

Zichermann and Cunningham's work targets marketers, application designers, and corporate brand and product managers. The authors demonstrate the ways in which gamification can be utilized in digital applications in order to acquire and engage consumers and users, shifting from traditional loyalty programs to engagement platforms. They define gamification as "the process of game-thinking and game mechanics to engage users and solve problems" (xiv). Zichermann and Cunningham outline areas of game fundamentals that focus on player motivation, game mechanics, design practices, and integration of social interactions. The book contains case studies of companies that apply gamification, as well as tutorials to develop game mechanics.

5. Game-Design Models and the Digital Economy

Beller, Jonathan. 2006. *The Cinematic Mode of Production: Attention Economy and the Society of the Spectacle.* **Hanover, NH: Dartmouth College Press.**

Beller posits new media (including cinema, television, video, computers, and the Internet) as the dominant mode of production in global, postindustrial capitalism. He argues that new media function as deterritorialized factories wherein spectators engage in value-productive labour. Beller explains that the commodification of experience and leisure time emerges because the exchange value of a commodity increases the more the commodity image gets consumed. Furthermore, the spectator or consumer performs the labour of a worker, beyond normal working hours, because watching becomes a productive labour act for which the spectator is paid in enjoyment. Beller provides numerous examples to demonstrate how this process takes place in current capitalist environments.

_____ **2006/07. "Paying Attention."** *Cabinet* **24: n.p. http://www.cabinetmagazine.org/issues/24/beller.php.**

Beller argues that attention is a commodity in the current neoliberal, global capitalist economy. In the twenty-first century media landscape, attention is constantly traded for information, whether in the form of media buyers in the advertising industry, in the entertainment economy (e.g., cinema, video games), or through content and information sharing in social networks. And attention is not only a commodity, but can be seen as productive labour, since attention produces capital. Using cinema as an example, Beller explains that the attention economy relies on the visual gaze and subsequent value production through the viewer; he describes this as a process wherein surplus value is extracted from spectators in deterritorialized factories that produce value for media companies. This process enables productive labour as well as the social cooperation necessary to maintain the capitalist hierarchy.

Bogost, Ian. 2007. *Persuasive Games: The Expressive Power of Videogames.* **Cambridge, MA: MIT Press.**

Bogost discusses his theory that video games are an expressive media making arguments through procedural rhetoric—the practice of persuasion through processes, and, in Bogost's case, computational processes in particular. According to Bogost, procedural computer representation differentiates itself from textual, visual, and plastic representation in that it is the only system in which process can be represented with process. He focuses on persuasive games, which he defines as "videogames that mount procedural rhetorics effectively" to influence players (46). Bogost reviews in detail the persuasive capabilities of video games in the realms of politics, advertising, and education from a theoretical and a game-design perspective.

_____ 2012. *Alien Phenomenology, or, What It's Like to Be a Thing.* **Minneapolis: University of Minnesota Press.**

Bogost proposes a form of study that goes beyond the way objects relate to humans. Rather than considering ideas as more valuable than things and our sense of being as the only way of being, Bogost suggests that we should begin to look at things through relations between object and object. In object-oriented ontology (OOO), things are at the centre of being, everything exists equally, and nothing (including humans) has special status. As an alternative term to OOO, Bogost suggests "unit operations." The term "unit" neither implies a subject nor requires materiality. Similarly, the term "operations" more accurately describes the processes in which all units behave and interact. Through the approaches of ontography (what reveals the object's existence and relations) and metaphorism (using metaphor to speculate about the unknowable), the phenomenology of units (or things or objects) can be studied, described, and analyzed while recognizing that we as humans cannot actually know what it means to be a thing. An OOO approach suggests a new form of humanism that does not rely on the correlational system of humans.

Dyer-Witheford, Nick, and Greig de Peuter. 2009. *Games of Empire: Global Capitalism and Video Games.* **Minneapolis: University of Minnesota Press.**

Dyer-Witheford and de Peuter argue that video games are a media of Empire—Michael Hardt and Antonio Negri's notion of a hypercapitalist sphere where the economic, cultural, and political issues of global capitalism take place in the same way as in the physical world. The authors' political critique relies on the concept that video games used to be primarily fun or pleasureful media, and now have been revealed to include facets of labour, authority, and capital. Drawing from Hardt and Negri, autonomist Marxism, and poststructuralist radicalism, Dyer-Witheford and de Peuter note the capitalist domination in video games in the form of network power, with multiple institutional agencies shaping and participating in the video game space. Virtual games are examples of Empire that highlight its constitution and conflicts, maintaining it and, at times, offering the space to challenge and rebel against it.

Grimes, Sara M., and Andrew Feenberg. 2009. "Rationalizing Play: A Critical Theory of Digital Gaming." *The Information Society* 25 (2): 105–18. doi:10.1080/01972240802701643.

Grimes and Feenberg propose their theory of socially rationalized games through an analysis of *World of Warcraft*. They suggest that the societal forms of motivation developing systemically out of MMOGs progressively diminish the playfulness associated with the discovery-based motivation intrinsic to these environments. Like Deterding et al. (2011), Grimes and Feenberg acknowledge their dependence on Caillois's distinction between ludus and paidia (1961) in developing their case for video games as systems of social rationality that change the experience of play through the forms of standardization that occur in their large-scale use.

Galloway, Alexander R. 2006. *Gaming: Essays on Algorithmic Culture*. Minneapolis: University of Minnesota Press.

Based on the argument that video games are actions, Galloway develops a four-part system that incorporates theoretical insights while treating video games as material objects, regarding them as an active and material medium. Following these assumptions, Galloway differentiates between machine actions (by the computer software and hardware) and operator actions (by the players). Furthermore, he recognizes that games are made up of diegetic space (the sphere of narrative action) and nondiegetic space (elements that are inside the game apparatus, but outside the distinct character and story world). Between these categories emerge four game actions that comprise Galloway's system: the diegetic machine act, the nondiegetic operator act, the diegetic operator act, and the nondiegetic machine act. Building on this structure, the essays provide examples of video games and other media and look at gaming practices to analyze video games as a cultural form that is actively played rather than read or watched.

Hayles, N. Katherine. 2007. "Hyper and Deep Attention: The Generational Divide in Cognitive Modes." *Profession:* **187–99.**
Hayles examines the differences in cognitive styles between deep attention and hyper attention. Deep attention, common in the humanities, concentrates on a single object for an extended period and ignores other stimuli. Hyper attention switches the focus of attention rapidly and requires stimulation. Rather than advocating for one or the other cognitive mode, Hayles calls for a change in education systems that allows for both types of attention. Hayles notes that hyper attention can still be focused on single activities for long periods of time, e.g., in video games. Video games, however, offer high levels of stimulation through the escalating series of rewards that players experience, as well as feelings of autonomy, competence, and relatedness. This, Hayles suggests, offers important insights for educators, especially in consideration of the digital space and how technology can be used in pedagogical environments. Hayles offers examples of possible approaches to show that critical interpretation and practices common in the humanities can be taught to and applied by all students, whether they are more comfortable with hyper attention or deep attention, if presented in the right way.

McGonigal, Jane. 2011. *Reality is Broken: Why Games Make Us Better and How They Can Change the World.* **New York: Penguin.**
McGonigal frames her argument under a bold statement that reality is broken, especially when compared to games. Drawing upon her own experiences as an independent game designer (see worldwithoutoil.org) and building on definitions of games and utopia from the work of Bernard Suits, McGonigal argues that the global ascendance of video games as a cultural form signals a purposeful escape from established societal structures. In McGonigal's view, video games are fulfilling genuine intrinsic human needs—teaching, inspiring, engaging, and building communities—in ways that reality is no longer able to. Games and game design are not just a pastime and a craft, but instead offer current ways of thinking and leading in order to effect real changes in the world. McGonigal contends that as reality is broken, video game designers must set out to recreate it.

Nakamura, Lisa. 2009. "Don't Hate the Player, Hate the Game: The Racialization of Labor in World of Warcraft." *Critical Studies in Media Communication* **26 (2): 128–44. doi:10.1080/15295030902860252.**
Nakamura analyzes the racialization of informational labor in MMOs generally and *World of Warcraft* specifically. Chinese player workers, discriminatingly called "Chinese gold farmers" in the player community, are racialized and dehumanized by other *World of Warcraft* players. Analyzing examples of

machinama that negatively present and attach Chinese player workers, such as the well-known machinama Ni Hao, Nakamura points out the many ways in which these user-generated videos produce racist narratives that rely on the game world and thus distance themselves from "real world" racism. Gold farming as a labour practice, Nakamura argues, reveals the reality of the exploitative digital economy and informationalized capitalism. Immaterial labour that often gets treated as play in fact becomes pure, real work for gold farmers who work 12-hour shifts in factory-like settings for incredibly low wages. These worker players do not have the opportunity to play the game that they are experts in. While other players have the opportunity to fully engage in the games as a leisure activity and even produce additional game-related content—such as the racist, dehumanizing machinama that Nakamura analyzes—for fun, player workers do not have the opportunity to engage with the game in such a way. Instead, they become disliked, racialized, discriminated non-player characters.

_____ 2013. "'Words With Friends': Socially Networked Reading on *Goodreads*." *PMLA* 128 (1): 238–43. https://lnakamur.files.wordpress. com/2013/04/nakamura-22words-with-friends22-pmla.pdf.
Nakamura considers the cultural shift toward electronic literature, noting the move from p-books (print books) to e-books, and asks how reading changes in digital environments. Rather than relying primarily on the hardware contexts of digital environments, digital reading follows social media in claiming a more service-based nature. Nakamura points out that books have always promoted forms of social networking, and she predicts a continuation of such social behavior in the current digital generation. Goodreads provides a highly developed example of what a social, digital reading environment can look like: it contains social networking elements (e.g., inbox, notifications, status ticker), links to other social networks, invitation generators to add friends, and an option to be used in the format of different apps. Bookshelves are public and reading data is shared, allowing for a variety of social forms of engagement. However, Nakamura notes that this also turns users into collectable objects; by participating in an environment like Goodreads, users share their data and become items in a database. Thus the reader becomes a labourer by engaging in activities that combine play and labour. Although Goodreads positions itself as a passive conduit that facilitates folksonomic creation and individual contribution, Nakamura argues that reading is a social, economic, and cultural activity that is never passive.

Schenold, Terry. 2011. "The 'Rattomorphism' of Gamification." *Critical Gaming Project*. https://depts.washington.edu/critgame/wordpress/2011/11/the-rattomorphism-of-gamification/.

Schenold offers a strong critique of gamification, using the notion of "rattomorphism" (as termed by Arthur Koestler and applied by Alfie Kohn) to describe the common rewards- and incentive-driven conditioning of users. While such an approach may be effective in the short term, Schenold likens it to "digital meth," arguing that the incentivized activities of gamification quickly become corrosive, and any form of attentiveness or creativity that the user may have been engaged in falls apart quickly. Finally, Schenold points out that there is no game layer, because games cannot merely be stripped down to assemblages of techniques. Instead, there are reward layers or feedback layers that may draw inspiration from games, but merely "address our inner rat, not our inner 'gamer'" (n.p.).

Scholz, Trebor, ed. 2013. *Digital Labor: The Internet as Playground and Factory*. New York: Routledge.

This collection of essays examines the current digital space as a labour site or factory, and what implications this structure—dominated by profit-driven, oligarchic owners—has on the digital worker today. The authors recognize a continuation of traditional economies in the digital space, which enables free labour that may not seem like labour at all. While the social web may appear free, users pay through their participation and with their data, ultimately being sold as the product that they also consume. This raises a question about the difference between work and play online: digital activities often cloud the differentiation between nonproductive leisure activity and productive work activity. Playbor (play/labor) is an aspect of the gift economy, where users perform labour for fun. Notably, in his essay "Considerations on a Hacker Manifesto" (69–75), McKenzie Wark cautions against the rhetoric of gamification, arguing that it is a simulation of the gift economy, since it extracts labour in the form of play within a reciprocal structure that is not driven by the players but instead by business requirements.

Suits, Bernard. 2005. *The Grasshopper: Games, Life and Utopia*. Introduction by Thomas Hurka. Peterborough, ON: Broadview Press.

This philosophical dialogue, originally published in 1978, has been recognized as among the underrated philosophical works of the twentieth century. The book suggests that philosopher Ludwig Wittgenstein's conception of games as sharing certain family resemblances is insufficiently clear. Suits conceives of playing a game as a voluntary attempt to overcome unnecessary obstacles. A game is comprised of a goal, means of achieving that goal, rules, and what Suits calls the "lusory attitude," or the acceptance by players of inefficient rules for reaching the goal.

Wark, McKenzie. 2007. *Gamer Theory*. Cambridge, MA: Harvard University Press.
Wark engages in a theoretical discourse about our everyday lives by discussing concepts of meaning, space, nuanced thinking, the work/play dichotomy, subjectivity, and resistance or social change through examples of video games. He regards the "real world" as divided into games, thus deeming it a gamespace that exists everywhere. Because of this spread of the gamespace, play has become work and work has become play. In order to engage in a critical theory of action, Wark presses for play from within the game against gamespace. Wark encourages an active approach to theory that overcomes social binaries such as work/play by engaging in gamer subjectivity in order to delve deeply into gamespace. Overall, Wark encourages a form of play in and against gamespace that engenders new concepts.

6. Game-Design Insights and Best Practices

Aarseth, Espen J. 2012. "A Narrative Theory of Games." In *Proceedings of the International Conference on the Foundations of Digital Games (FDG '12)*, 129–33. New York: ACM. doi:10.1145/2282338.2282365.
Aarseth considers the foundational debate that took place in game studies between narratologists, who followed Janet Murray in approaching video games and electronic texts as stories, and ludologists, who contended with Jesper Juul that the computer game is not simply a narrative medium. Aarseth sees video games as a combination of games and stories through software, one that can result in a variety of ludo-narratological constructs. This ludo-narrative design space consists of four dimensions: world, objects, agents, and events. Aarseth sees agents/characters as the most important dimension in video games, a key element differentiating video games from other narrative environments.

Anthropy, Anna. 2012. *Rise of the Videogame Zinesters: How Freaks, Normals, Amateurs, Artists, Dreamers, Dropouts, Queers, Housewives, and People Like You Are Taking Back an Art Form*. New York: Seven Stories Press.
Anthropy calls for more people to make video games in order to broaden the perspectives communicated through video games and thus push against the exclusive nature of current video game culture. She argues that the current video game scene and its history is dominated by a small group of people—educated men who have grown up playing games and then decided

to become game designers. Because of this, most games communicate stories and experiences from that male perspective, and games lack diversity. Anthropy argues that since games are particularly good at exploring dynamics, relationships, and systems, they are experiences created by rules. As the player must play the game in order for it to take place, it is through the player interaction with the rules that it becomes a game. Based on this requirement for interaction, the game creator tells stories not just through the content, but also through the design and the system of the game. Highly personal, complex stories can be told in this way, which is why Anthropy highlights the importance of bringing in more perspectives. In order to facilitate this, Anthropy describes different forms of hacking, "modding" (modifying), and game development that do not require any coding knowledge or particular design skills. Game-design tools are becoming increasingly available and accessible to wider audiences; as such, Anthropy calls for the rise of video game zinesters—hobbyists, makers, and players—to express their stories in the form of video games.

Bjork, Staffan, and Jussi Holopainen. 2005. *Patterns in Game Design.* Hingham, MA: Charles River Media.
Bjork and Holopainen outline an approach to game design that considers elements of games as game-design patterns that can be analyzed and applied. This toolset offers game designers and scholars a language to talk about the elements of gameplay, which is currently lacking. Bjork and Holopainen explain that design patterns are useful for analytical purposes of existing games or prototypes and for game design during the creation of games, since they can help at the stage of idea generation and structure the development of game concepts. The authors aim to construct a language based on interactions, rather than narratology, as has been common in game studies in the past.

Bogost, Ian. 2011. *How To Do Things With Videogames.* Minneapolis: University of Minnesota Press.
Bogost provides an overview of the many different applications of video games. As he demonstrates, combinations of applications reveal that the medium of video games is much broader, richer, and more relevant than generally acknowledged. The extensive scope of video games indicates that they should not be simplified and regarded as a medium for leisure or productivity, but recognized as a medium that offers a wide range of potential uses.

Caillois, Roger. (1961) 2001. *Man, Play, and Games.* Translated by Meyer
    Barash. New York: Free Press of Glencoe. Reprint, Chicago: Univer-
    sity of Illinois Press.

Caillois assesses social practices as rule-bound games that serve to limit freer
forms of play within cultures. Structures of games culturally acknowledged
as such (e.g., chess) derive from outmoded social practices. Caillois's work
on games has been particularly significant in defining play and games. He
defines gameplay as that which is free, separate, uncertain, unproductive,
governed by rules, and make-believe. Furthermore, Caillois argues that
all games contain one or a combination of the following categories: agon
(competition), alea (chance), mimicry (simulation), and ilinx (vertigo). The
distinction between active, exuberant, and spontaneous "paidia" and calcu-
lating, contrived "ludus" is still relied on and often referenced by contemporary game scholars.

Coleman, Susan L., Ellen S. Menaker, Jennifer McNamara, and Tristan
    E. Johnson. 2015. "Communication for Stronger Learning Game
    Design." In *Design and Development of Training Games,* edited by Talib
    S. Hussain and Susan L. Coleman. New York: Cambridge University
    Press, 31–54. doi:10.1017/CBO9781107280137.003.

Coleman, Menaker, McNamara, and Johnson discuss the challenges of
learning game design (LGD) teams, and how to improve their efficiency
and effectiveness through communication. They argue that a shared men-
tal model is needed in order to address the challenge of diversity in LGD
teams. The authors conclude that one's team must understand the individual expectations and principles of team members in order to promote
communication well.

Deterding, Sebastian. 2012. "Gamification: Designing for Motivation."
    Interactions 19 (4): 14–17. doi:10.1145/2212877.2212883.

This forum offers multiple perspectives relevant to the discourse on gamification by Sebastian Deterding, Judd Antin, Elizabeth Lawley, and Rajat Paha-
ria. Antin asserts that online gamification participants do not work for free,
but are paid with good feelings. Gamification mechanisms such as badges
have a bad reputation, not because they do not work, but because they are
frequently implemented inappropriately for the audience and purpose. As
Lawley points out, successful gamification applies game design, not solely
game components. The forum urges practitioners to recognize the value of
gamification beyond the stock features commonly implemented.

_____ 2015. "The Lens of Intrinsic Skill Atoms: A Method for Gameful Design." *Human-Computer Interaction* 30 (3–4): 294–335. doi :10.1080/07370024.2014.993471.

Deterding analyzes conceptually the design challenges behind reviewing existing methods of gameful design. He presents a method (the "Lens of Intrinsic Skill Atoms") that restructures the challenges inherent in the end user's goal pursuit into a system that would provide enjoyable and motivating experiences. Deterding works with two case studies to illustrate his method: "Innovating Gameful Design" and "Evaluating Gameful Design." He concludes with a list of criteria for gameful design, shows how his proposed method meets them all, and affirms the usefulness of his approach.

Ferrara, John. 2012. *Playful Design: Creating Game Experiences in Everyday Interfaces.* **Brooklyn, NY: Rosenfeld Media.**

Ferrara structures his book as a guide for user experience (UX) designers to apply game design as part of their approach. While critical of the buzz around gamification and the imprecise application of the term, Ferrara stresses that game-design approaches can be highly successful if focused on the player experience. He offers an extensive overview introducing the reader to game-design approaches that may be relevant to general UX design. The first section, "Playful Thinking," explains the ways in which games can be effective when applied to the everyday or the real world, defines games and their relation to everyday experiences, and outlines aspects of player experience and player motivation. "Designing Game Experiences" addresses more practical aspects of building user experiences based on game-design approaches. This section outlines tips for building game concepts, creating prototypes, play testing, behavioural tools, and the potential of rewards in games. The final section, "Playful Design in User Experience," looks in more detail at how games can be used as methods for action, learning, and persuasion in the everyday. Ferrara concludes with speculations on future trends.

Gamification Wiki. n.d. "Gamification." https://badgeville.com/wiki/ gamification. **Dublin, CA: Badgeville.**

This wiki offers an array of resources related to gamification and game mechanics. It contains general information on gamification as well as links to books, examples, presentations, and videos. Specific areas of gamification include education, marketing, government, social good, and design.

Høgenhaug, Peter Steen. 2012. "Gamification and UX: Where Users Win or Lose." *Smashing Magazine.* **n.p. http://uxdesign.smashingmagazine. com/2012/04/26/gamification-ux-users-win-lose/.**
Høgenhaug outlines the ways in which gamification can improve the user experience of websites and applications. He begins by defining four key actions that comprise games: play, pretending, rules, and goals. Practitioners who plan to use gamification should not consider it an add-on, but include it in the design process itself. Game models and approaches that work well in UX design include tangible user interfaces, constructive and helpful feedback, storytelling, and Easter eggs. Gamification should not be overused, but rather considered a tool to improve user experience by complementing the content and structure of a site or app. Høgenhaug also suggests what to avoid when using gamification.

Kim, Bohyun. 2012. "Harnessing the Power of Game Dynamics: Why, How to, and How Not to Gamify the Library Experience." *College & Research Libraries* **News 73 (8): 465–69.**
Kim acknowledges that gamification of the library experience is becoming increasingly common in academic libraries. She recognizes the advantages of gamification in terms of motivation, engagement, and increased achievements of tasks toward a goal. Kim also outlines tactical opportunities and approaches to avoid when gamifying the library experience.

Liu, Yefeng, Todorka Alexandrova, and Tatsuo Nakajima. 2011. "Gamifying Intelligent Environments." **In** *Proceedings of the 2011 International ACM Workshop on Ubiquitous Meta User Interfaces (Ubi- MUI '11),* **7–12. New York: ACM. doi:10.1145/2072652.2072655.**
Liu, Alexandrova, and Nakajima review the ways in which digital designers apply gamification methods in the design of intelligent environments in order to improve user engagement. They provide two case studies to determine the effectiveness of this approach (UbiAsk, a crowdsourcing application, and EcoIsland, a persuasive application to reduce carbon dioxide emissions). The authors conclude that gamification approaches are only effective in driving participation when they are implemented as additional components supporting an otherwise functioning app or environment. Game actions also must be initiated by a deeper game structure throughout the environment.

McGonigal, Jane. 2008. "Engagement Economy: The Future of Massively Scaled Collaboration and Participation." Edited by Jess Hemerly and Lisa Mumbach. Palo Alto, CA: Institute for the Future, Technology

Horizons Program. **http://www.iftf.org/uploads/media/Engage-ment_Economy_sm_0.pdf.**

McGonigal contends that the current economy of engagement is no longer just about competing for attention; now, engagement relies on interaction and contribution by users. She claims that innovative organizations need to tackle the challenge of participation bandwidth, and ought to learn from the world of play to do so. McGonigal explains that the digital environment contains more and more mass collaboration and crowdsourcing platforms and networks, which makes it increasingly difficult to encourage and maintain engagement. She asserts that gaming approaches can help to optimize participation bandwidth because of the importance of emotional incentives in today's social mindset. McGonigal infers that designing for positive emotional goals will keep users of all levels of participation more engaged. Finally, she suggests that the most effective way of ensuring a continuous engagement life cycle is to structure platforms that empower community members to invent their own tasks.

Play the Past. **2010. http://www.playthepast.org.**

Play the Past is a collaboratively authored and edited website that looks at the intersections between cultural heritage and all kinds of games. The authors write about diverse topics related to culture and games, including theoretical approaches, philosophical reflections, and practical considerations. Topics range from online gaming experiences to pedagogical applications of games.

Salen, Katie, and Eric Zimmerman. 2004. *Rules of Play: Game Design Fundamentals.* **Cambridge, MA: MIT Press.**

Salen and Zimmerman analyze games as designed systems, and outline key concepts for the creation of games, thus establishing a critical discourse for game design. The authors define core concepts (such as play, games, design, systems, and interactivity), as well as rules, rule levels, and rule systems. As Salen and Zimmerman explain, all games have rules, and the rules of a game are what distinguish it from other games. Thus, players accept the rules and limitations defined by a particular game when they play it. Salen and Zimmerman note that the play of a game is the experiential aspect of a game, and outline different categories of play type, as well as three phenomena of play behaviour (game play, ludic activities, and being playful). The authors also outline the social relationships, player roles, and community aspects of gameplay, as well as the structure, environment, and social contracts that are required for the culture of a game to flourish.

7. A Complete Alphabetical List of Selections

Aarseth, Espen J. 1997. "Introduction." In *Cybertext: Perspectives on Ergodic Literature*, 1–23. Baltimore: Johns Hopkins University Press.

_____ 2012. "A Narrative Theory of Games." In *Proceedings of the International Conference on the Foundations of Digital Games (FDG '12)*, 129–33. New York: ACM. doi:10.1145/2282338.2282365.

Anthropy, Anna. 2012. *Rise of the Videogame Zinesters: How Freaks, Normals, Amateurs, Artists, Dreamers, Dropouts, Queers, Housewives, and People Like You Are Taking Back an Art Form.* New York: Seven Stories Press.

Balsamo, Anne. 2011. Introduction: "Taking Culture Seriously in the Age of Innovation." In *Designing Culture: The Technological Imagination at Work*, 2–25. Durham, NC: Duke University Press.

Beller, Jonathan. 2006. *The Cinematic Mode of Production: Attention Economy and the Society of the Spectacle.* Hanover, NH: Dartmouth College Press.

_____ 2006/07. "Paying Attention." *Cabinet* 24: n.p. http://www.cabinetmagazine.org/issues/24/beller.php.

Bjork, Staffan, and Jussi Holopainen. 2005. *Patterns in Game Design.* Hingham, MA: Charles River Media.

Blizzard Entertainment, Inc. 2005. *World of Warcraft* (WoW) [video game]. https://worldofwarcraft.com/cn-us/.

Bogost, Ian. 2007. *Persuasive Games: The Expressive Power of Videogames.* Cambridge, MA: MIT Press.

_____ 2011a. *How To Do Things With Videogames.* Minneapolis: University of Minnesota Press.

_____ 2011b. "Persuasive Games: Exploitationware" [blog post]. *Gamasutra.* http://www.gamasutra.com/view/feature/6366/persuasive_games_exploitationware.php.

_____ 2012. *Alien Phenomenology, or, What It's Like to Be a Thing.* Minneapolis: University of Minnesota Press.

Caillois, Roger. (1961) 2001. *Man, Play, and Games.* Translated by Meyer Barash. New York: Free Press of Glencoe. Reprint, Chicago: University of Illinois Press.

Carson, Stephen, and Jan Philipp Schmidt. 2012. "The Massive Open Online Professor." *Academic Matters: The Journal of Higher Education*, n.p. http://www.academicmatters.ca/2012/05/the-massive-open-online-professor/.

CCP Games. 2003. *EVE Online* [video game]. http://www.eveonline.com/.

Center for Interdisciplinary Inquiry and Innovation in Sexual and Reproductive Health (Ci3), University of Chicago. 2013. *The Source* [video game]. https://ci3.uchicago.edu/portfolio/the-source/.

Chamberlin, Barbara, Jesús Trespalacios, and Rachel Gallagher. 2014. "Bridging Research and Game Development: A Learning Games Design Model for Multi-Game Projects." In *Educational Technology Use and Design for Improved Learning Opportunities*, edited by Mehdi Khosrow-Pour, 151–71. Hershey, PA: IGI Global. doi:10.4018/978-1-4666-6102-8.ch008.

Chang, Edmond. 2012. "Video+Game+Other+Media: Video Games and Remediation" [blog post]. *Critical Gaming Project.* https://depts.washington.edu/critgame/wordpress/2012/01/videogameothermedia-video-games-and-remediation/.

Clement, Tanya. 2011. "Knowledge Representation and Digital Scholarly Editions in Theory and Practice." *Journal of the Text Encoding Initiative* 1: n.p. doi:10.4000/jtei.203.

Coleman, Susan L., Ellen S. Menaker, Jennifer McNamara, and Tristan E. Johnson. 2015. "Communication for Stronger Learning Game Design." In *Design and Development of Training Games*, edited by Talib S. Hussain and Susan L. Coleman. New York: Cambridge University Press, 31–54. doi:10.1017/CBO9781107280137.003.

Crowley, Dennis, and Naveen Selvadurai. 2009. *Foursquare.* New York: Foursquare. https://foursquare.com.

Cyber Creations Inc. 2002. *MMORPG.com* [video game website]. http://www.mmorpg.com/.

Danforth, Liz. 2011. "Gamification and Libraries." *Library Journal* 136 (3): 84–85.

Davidson, Cathy N. 2011. "Why Badges? Why Not?" [blog post]. *HASTAC.* https://www.hastac.org/blogs/cathy-davidson/2011/09/16/why-badges-why-not.

Davidson, Cathy N., and David Theo Goldberg. 2004. "Engaging the Humanities." *Profession*: 42–62. doi:10.1632/074069504X26386.

De Carvalho, Carlos Rosemberg Maia, and Elizabeth S. Furtado. 2012. "Wikimarks: An Approach Proposition for Generating Collaborative, Structured Content from Social Networking Sharing on the Web." In *Proceedings of the 11th Brazilian Symposium on Human Factors in Computing Systems (IHC '12)*, 95–98. Porto Alegre, Brazil: Brazilian Computer Society.

De Paoli, Stefano, Nicolò De Uffici, and Vincenzo D'Andrea. 2012. "Designing Badges for a Civic Media Platform: Reputation and Named Levels." In *Proceedings of the 26th Annual BCS Interaction Specialist Group Conference on People and Computers (BCS-HCI '12)*. Swinton, UK: British Computer Society, 59–68.

Deterding, Sebastian. 2012. "Gamification: Designing for Motivation." *Interactions* 19 (4): 14–17. doi:10.1145/2212877.2212883.

_____ 2015. "The Lens of Intrinsic Skill Atoms: A Method for Gameful Design." *Human-Computer Interaction* 30 (3–4): 294–335. doi:10.1080/0737 0024.2014.993471.

Deterding, Sebastian, Dan Dixon, Rilla Khalad, and Lennart E. Nacke. 2011. "From Game Design Elements to Gamefulness: Defining 'Gamification.'" In *Proceedings of the 15th International Academic MindTrek Conference: Envisioning Future Media Environments (MindTrek '11)*, 9–15. New York: ACM. doi:10.1145/2181037.2181040.

Dickey, Michele D. 2007. "Game Design and Learning: A Conjectural Analysis of How Massively Multiple Online Role-Playing Games (MMORPGs) Foster Intrinsic Motivation." *Educational Technology Research and Development* 55 (3): 253–73.

Douma, Michael. 2011. "What is Gamification?" [blog post]. *Idea*. http://www.idea.org/blog/2011/10/20/what-is-gamification/.

Drucker, Johanna. 2006. "Graphical Readings and the Visual Aesthetics of Textuality." *TEXT: An Interdisciplinary Annual of Textual Studies* 16 267–76. http://www.jstor.org/stable/30227973.

_____ 2011a. "Humanities Approaches to Graphical Display." *Digital Humanities Quarterly* 5 (1): n.p. http://www.digitalhumanities.org/dhq/vol/5/1/000091/000091.html.

_____ 2011b. "Humanities Approaches to Interface Theory." *Culture Machine* 12: 1–20. http://www.culturemachine.net/index.php/cm/article/viewArticle/434.

Drucker, Johanna, and Jerome McGann. 2000. *Ivanhoe.* SpecLab (University of Virginia). http://www.ivanhoegame.org/?page_id=21.

Dyer-Witheford, Nick, and Greig de Peuter. 2009. *Games of Empire: Global Capitalism and Video Games.* Minneapolis: University of Minnesota Press.

Ferrara, John. 2012. *Playful Design: Creating Game Experiences in Everyday Interfaces.* Brooklyn, NY: Rosenfeld Media.

Galloway, Alexander R. 2006. *Gaming: Essays on Algorithmic Culture.* Minneapolis: University of Minnesota Press.

Galloway, Alexander R., Carolyn Kane, Adam Parrish, Daniel Perlin, DJ / rupture and Matt Shadetek, Mushon Zer-Aviv, and the RSG Collective. *Kriegspiel* [video game]. New York University. http://r-s-g.org/kriegspiel/index.php.

Gamification Wiki. "Gamification." https://badgeville.com/wiki/gamification. Dublin, CA: Badgeville.

Gibson, David, Clark Aldrich, and Marc Prensky, eds. 2007. *Games and Simulations in Online Learning: Research and Development Frameworks.* Hershey, PA: Information Science Publishing.

Graham, Adam. 2012. "Gamification: Where's the Fun in That?" [blog post]. *Campaign* 43: 47. http://www.campaignlive.co.uk/article/1156994/gamification-wheres-fun-that.

Grimes, Sara M., and Andrew Feenberg. 2009. "Rationalizing Play: A Critical Theory of Digital Gaming." *The Information Society* 25 (2): 105–18. doi:10.1080/01972240802701643.

Groh, Fabian. 2012. "Gamification: State of the Art Definition and Utilization." In *Proceedings of the 4th Seminar on Research Trends in Media Informatics (RTMI '12),* edited by Naim Asaj, et al. Ulm, Germany: Institute of Media Informatics, Ulm University, 39–46. http://hubscher.org/roland/courses/hf765/readings/Groh_2012.pdf.

Guldi, Jo. 2013. "Reinventing the Academic Journal." In *Hacking the Academy: New Approaches to Scholarship and Teaching from Digital Humanities,* edited by Daniel J. Cohen and Tom Scheinfeldt, 19–24. Ann Arbor: University of Michigan Press. doi:10.3998/dh.12172434.0001.001.

Hayles, N. Katherine. 2007. "Hyper and Deep Attention: The Generational Divide in Cognitive Modes." *Profession*: 187–99.

_____ 2008. *Electronic Literature: New Horizons for the Literary.* Notre Dame, IN: University of Notre Dame Press.

Høgenhaug, Peter Steen. 2012. "Gamification and UX: Where Users Win or Lose." *Smashing Magazine.* n.p. http://uxdesign.smashingmagazine. com/2012/04/26/gamification-ux-users-win-lose/.

Huizinga, Johan. 1949. *Homo Ludens: A Study of the Play-element in Culture.* London: Routledge and Kegan Paul.

Insemtives. 2009–13. *INSEMTIVES.* http://insemtives.eu/.

Jagoda, Patrick. 2013. "Gamification and Other Forms of Play." *Boundary 2* 40 (2): 113–44. doi:10.1215/01903659–2151821.

_____ 2014. "Gaming the Humanities." *Differences* 25 (1): 189–215.

Jakobsson, Mikael. 2011. "The Achievement Machine: Understanding Xbox 360 Achievements in Gaming Practices." *Game Studies: The International Journal of Computer Game Research* 11 (1): n.p. http://gamestudies. org/1101/articles/jakobsson.

Jensen, Matthew. 2012. "Engaging the Learner." *Training and Development* 66 (1): 40.

Jing, Tee Wee, Yue Wong Seng, and Raja Kumar Murugesan. 2015. "Learning Outcome Enhancement via Serious Game: Implementing Game-Based Learning Framework in Blended Learning Environment." *5th International Conference on IT Convergence and Security (ICITCS)*, 24–27 August. doi:10.1109/ICITCS.2015.7292992.

Jones, Steven E. 2009. "Second Life, Video Games, and the Social Text." *PMLA* 124 (1): 264–72.

_____ 2011. "Performing the Social Text: Or, What I Learned from Playing Spore." *Common Knowledge* 17 (2): 283–91. doi:10.1215/0961754X-1187977.

_____ 2013. *The Emergence of the Digital Humanities.* London and New York: Routledge.

Kapp, Karl M. 2012. *The Gamification of Learning and Instruction: Game-Based Methods and Strategies for Training and Education.* San Francisco: Pfeiffer.

Kim, Bohyun. 2012. "Harnessing the Power of Game Dynamics: Why, How to, and How Not to Gamify the Library Experience." *College & Research Libraries News* 73 (8): 465–69.

Kopas, Merritt. 2012. *Lim* [video game]. New York: Games for Change. http://www.gamesforchange.org/play/lim/.

Latour, Bruno. 2009. "A Cautious Prometheus? A Few Steps Towards a Philosophy of Design (with Special Attention to Peter Sloterdijk)." In *Networks of Design: Proceedings of the 2008 Annual International Conference of the Design History Society,* edited by Fiona Hackne, Jonathan Glynne, and Viv Minto, 2–10. Boca Raton, FL: Universal Publishers.

Liu, Yefeng, Todorka Alexandrova, and Tatsuo Nakajima. 2011. "Gamifying Intelligent Environments." In *Proceedings of the 2011 International ACM Workshop on Ubiquitous Meta User Interfaces (Ubi-MUI '11),* 7–12. New York: ACM. doi:10.1145/2072652.2072655.

Losh, Elizabeth. 2012. "Hacktivism and the Humanities: Programming Protest in the Era of the Digital University." In *Debates in the Digital Humanities,* edited by Matthew K. Gold, 161–86. Minneapolis: University of Minnesota Press. http://dhdebates.gc.cuny.edu/debates/text/32.

Maxis. 2008. *Spore* [video game]. Redwood City, CA: Electronic Arts Inc. http://www.spore.com/.

Maxis and The Sims Studio. 2000/2006–. *The Sims* [video game]. Redwood City, CA: Electronic Arts Inc. http://www.thesims.com/en-us.

McGann, Jerome. 2001. *Radiant Textuality: Literature after the World Wide Web.* New York: Palgrave.

———— 2005. "Like Leaving The Nile. IVANHOE, A User's Manual." *Literature Compass* 2: 1–27. http://www2.iath.virginia.edu/jjm2f/old/compass.pdf.

McGonigal, Jane. 2008. "Engagement Economy: The Future of Massively Scaled Collaboration and Participation." Edited by Jess Hemerly and Lisa Mumbach. Palo Alto, CA: Institute for the Future, Technology Horizons Program. http://www.iftf.org/uploads/media/Engagement_Economy_sm_0.pdf.

_____ 2011. *Reality is Broken: Why Games Make Us Better and How They Can Change the World.* New York: Penguin.

Meier, Sid. 1991. *Civilization* [video game]. Hunt Valley, MD: MicroProse. http://www.civilization.com.

Mojang and Microsoft Studios. 2011. *Minecraft* [video game]. Stockholm: Mojang; Redmond, WA: Microsoft Studios. https://minecraft.net.

Multiplayer Online Battle Arena (MOBA) games. 2009–.

Mysirlaki, Sofia, and Fotini Paraskeva. 2012. "Leadership in MMOGs: A Field of Research on Virtual Teams." *Electronic Journal of E-Learning* 10 (2): 223–34.

Nakamura, Lisa. 2009. "Don't Hate the Player, Hate the Game: The Racialization of Labor in World of Warcraft." *Critical Studies in Media Communication* 26 (2): 128–44. doi:10.1080/15295030902860252.

_____ 2013. "'Words With Friends': Socially Networked Reading on Goodreads." *PMLA* 128 (1): 238–43. https://lnakamur.files.wordpress.com/2013/04/nakamura-22words-with-friends22–pmla.pdf.

Play the Past. 2010. http://www.playthepast.org.

Polytron Corporation. 2012. *Fez* [video game]. http://fezgame.com/.

Ramsay, Stephen, and Geoffrey Rockwell. 2012. "Developing Things: Notes Toward an Epistemology of Building in the Digital Humanities." In *Debates in the Digital Humanities*, edited by Matthew K. Gold, 75–84. Minneapolis: University of Minnesota Press. http://dhdebates.gc.cuny.edu/debates/part/3.

Real-Time Strategy Games (RTSGs). 1982–.

Ritterfeld, Ute, Michael Cody, and Peter Vorderer, eds. 2009. *Serious Games: Mechanisms and Effects.* New York and London: Routledge.

Rockstar Games, Inc. 1997. *Grand Theft Auto* (GTA) [video game]. http://www.rockstargames.com/grandtheftauto/.

Rockwell, Geoffrey. 2003. "Serious Play at Hand: Is Gaming Serious Research in the Humanities?" *TEXT Technology* 2: 89–99.

Rose, Frank. 2011. *The Art of Immersion: How the Digital Generation is Remaking Hollywood, Madison Avenue, and the Way We Tell Stories.* New York: W.W. Norton.

Ryan, Marie-Laure. 1994. "Immersion vs. Interactivity: Virtual Reality and Literary Theory." *Postmodern Culture* 5 (1): 110–37. doi:10.1353/sub.1999.0015.

Salen, Katie, and Eric Zimmerman. 2004. *Rules of Play: Game Design Fundamentals.* Cambridge, MA: MIT Press.

Schenold, Terry. 2011. "The 'Rattomorphism' of Gamification." Critical Gaming Project. n.p. https://depts.washington.edu/critgame/wordpress/2011/11/the-rattomorphism-of-gamification/.

Scholz, Trebor, ed. 2013. *Digital Labor: The Internet as Playground and Factory.* New York: Routledge.

Squire, Kurt. 2008. "Open-Ended Video Games: A Model for Developing Learning for the Interactive Age." In *The Ecology of Games: Connecting Youth, Games, and Learning*, edited by Katie Salen, 167–98. Cambridge, MA: MIT Press.

Stack Exchange Network. 2013. *Stack Overflow.* http://stackoverflow.com.

Suits, Bernard. 2005. *The Grasshopper: Games, Life, and Utopia.* Introduction by Thomas Hurka. Peterborough, ON: Broadview Press.

Van Staalduinen, Jan-Paul, and Sara de Freitas. 2011. "A Game-Based Learning Framework: Linking Game Design and Learning Outcomes." In *Learning to Play: Exploring the Future of Education with Video Games,* edited by Myint Swe Khine, 29–54. New York: Peter Lang.

Vetch, Paul. 2010. "From Edition to Experience: Feeling the Way towards User-Focused Interfaces." In *Electronic Publishing: Politics and Pragmatics,* edited by Gabriel Egan, 171–84. New Technologies in Medieval and Renaissance Studies 2. Tempe, AZ: Iter Inc., in collaboration with the Arizona Center for Medieval and Renaissance Studies.

Wark, McKenzie. 2007. *Gamer Theory.* Cambridge, MA: Harvard University Press.

Werbach, Kevin. 2014. "(Re)Defining Gamification: A Process Approach." In *Proceedings of the 9th International Conference on Persuasive Technology*

(PERSUASIVE 2014), Lecture Notes in Computer Science 8462: 266–72. doi:10.1007/978-3-319-07127-5_23.

Zichermann, Gabe, and Christopher Cunningham. 2011. *Gamification By Design: Implementing Game Mechanics in Web and Mobile Apps.* Sebastopol, CA: O'Reilly Media.

Zynga Inc. 2009. *FarmVille* [video game]. Facebook and HTML 5. http://company.zynga.com/games/farmville.

III. Social Knowledge Creation Tools

The methods and channels for social knowledge creation proliferate alongside an increasingly networked world. Individuals, corporations, academic organizations, and others have all developed and employed tools of varying usefulness and relevance for social knowledge creation. The following section of the annotated bibliography outlines a brief scan of current digital social knowledge creation tools. By collecting these diverse tools into a single compiled list, we attempt to describe the breadth of social knowledge creation applications and services available. From the commercial to the open source, the proprietary to the freely available, these tools all contribute to social knowledge creation in the digital sphere at large.

Certain resources included here remain specific to the digital humanities community, while many other examples fall outside of that delineated space. Frequently, the latter can be applied or repurposed for the former. With this potential usage in mind, we include selections that may, at first glance, appear more or less relevant than others. Theoretically, an abundance of digital applications and services could be classified as social knowledge creation tools. In order to present a manageable amount of relevant information, we have divided the selections of this annotated bibliography into 5 categories of 55 individual entries, accompanied by a complete alphabetical list:[10]

1. Collaborative Annotation

2. User-Derived Content

3. Folksonomy Tagging

4. Community Bibliography

5. Shared Text Analysis

6. A Complete Alphabetical List of Selections

Of note, we have included only tools that were active at the time of writing. The temporal nature of the Internet dictates that many of these tools will eventually become obsolete. Instead of considering the compilation below

---

[10] Please note that cross-posted entries are marked with an asterisk (*) after the first instance.

as an authoritative, static list, therefore, we encourage readers to consider the included selections as an archival snapshot of current social knowledge creation tools. We hope that this list may serve as a representative of early twenty-first-century social knowledge creation tools, even as they morph and change with Internet trends and technology.

The outlined five categories intentionally compliment each other, and often a multipurpose or easily extensible entry may relate to multiple categories. While some of the tools are purposely dedicated to social knowledge creation, others can be applied for use in a social knowledge creation context, or can be hacked or repurposed to serve social knowledge creation ends. The first category, "Collaborative Annotation," features tools that facilitate multi-participant annotation of a shared document, image, or other digital artifact. Annotation is pivotal to scholarly research and production. Remediating annotation practices has been a pressing concern as an increasing amount of scholarly resources and projects move into the online sphere. Furthermore, the rise of social knowledge creation practices has encouraged the active development of collaborative annotation—the practice of annotating a document along with a group of online collaborators. Of course, there is no one right way to engage in collaborative annotation. This practice is also not limited to the academy; in fact, collaborative annotation tools have been largely taken up in the project management and business world, where many teams jointly develop and comment on documents or prototypes. The 17 tools in this category have been selected based on their relevance, usability, portability, and overall capacity to instigate social knowledge creation via shared annotation. Although the predominant focus of this category is concerned with how collaborative annotation can induce social knowledge creation in the scholarly community, tools that are relevant to various communities and can be applied broadly present perhaps the most interesting opportunities for initiating truly *social* knowledge creation.

"User-Derived Content," the second constellation of entries, comprises tools and services that foster the development of user content. Online repositories that encourage the production of user-derived content showcase the breadth of and possibilities for social knowledge creation in the digital realm. Although individuals have been generating content (read: interacting, making artifacts, sharing experiences) for centuries, the Internet has facilitated the creation of vastly popular, widespread, and specifically delineated spaces for presenting this content. Issues arise as this content is farmed or otherwise exploited by corporations, many of which actively promote the creation of user-generated or user-derived content. The tools and services highlighted here take a different tack from those of their more boldly capitalist digital brethren. The 10 selections comprise exhibits, databases, networks, and game-based credential systems that facilitate social knowledge production by the very nature of their form. Many of these tools are for use in an academic or otherwise educational context. Often, these tools and services enable users to both generate content and manipulate, catalogue, visualize, or otherwise engage with their own and others' content.

The third category, "Folksonomy Tagging," includes tools and services for folksonomy development via content producer and consumer tagging. Through folksonomy or social tagging practices, individuals can add metadata to artifacts for their own or others' searching and indexing benefit. Folksonomy tagging creates an infrastructure of navigable digital images, texts, videos, and sites, and provokes social knowledge creation by supplying the tools to efficiently access and otherwise manipulate user-generated content. Although folksonomy tagging is most common in social networks, this category includes six diverse selections that range from predominantly social media sites to digital bookmarking applications to community commerce spaces. The variance between entries speaks to the many ways in which folksonomy tagging can be used to foster social knowledge creation.

"Community Bibliography" describes tools and applications that enable collaborative and shared cataloguing and reference management. A variety of cataloguing and reference management systems and resources have been developed to aid scholars in the creation, organization, application, and publication of bibliographies. This category includes 15 browser-based, desktop, and command-line tools. In addition to providing means to a more efficient workflow process through simple import and export functions, many of these tools also allow for easier methods of publication or creation of online exhibits. We expand the concept of community bibliography to include comprehensive code repositories, pivotal as they are for organizing, accessing, and harnessing contemporary social knowledge creation. Online reference management and social bookmarking systems are increasingly structured as social networks or in ways that encourage collaboration by allowing for shared lists, libraries, notes, and discussion forums. Many tools also offer tagging functions in a folksonomy style to allow for higher searchability and dynamic recommendations of sources based on similar users. The majority of tools listed in this category target an academic audience, with certain selections geared toward humanities scholars and others toward scientists.

The final category, "Shared Text Analysis," outlines web-based tools designed for collaborative text analysis and visualization. Increasingly, literary scholars recognize computer-aided text analysis as a relevant method for humanities work. Additionally, a growing number of online tools create new opportunities for sharing and collaboration during the text analysis process. This category outlines seven web-based tools and applications that supplement scholarly work in the realms of textual analysis, text comparison, annotation, markup, tagging, and visualization. The online nature of the tools makes collaborative work easier for textual scholars, as multiple users can view, access, and work on the same texts.

Social Knowledge Creation Tools is intended to round out the larger environmental scan of current academic, para-academic, and non-academic instantiations and explorations of social knowledge creation. As social knowledge creation and the digital environment become increasingly intertwined, it is important to examine who is involved in the shaping of this field, and how. Ideally, the reader of this section of the annotated bibliography will benefit from the breadth and depth of selected tools, services, and applications, and find it a useful resource for the active study, participation, and instigation of social knowledge creation.

1. Collaborative Annotation

Abilian SAS. 2010–15. *co-ment.* **http://www.co-ment.com/.**
co-ment is a web service for viewing, creating, and interacting with annotations. With co-ment, a user may upload or create texts online, invite designated users to comment on files, and revise drafts. According to its website, co-ment is "the reference Web service for submitting texts to comments and annotations." Plug-ins for multiple content management systems and platforms can be created using an API. Notably, co-ment is open source and web-based.

Diigo, Inc. *Diigo.* **2017. https://www.diigo.com/.**
Launched in 2006, Diigo may be best conceived of as a platform for collecting and managing research (including text, bookmarks, images, and documents). With a professed focus on enhancing the e-reading experience, Diigo enables a variety of online practices, from social bookmarking to comprehensive search to multi-user annotation. This service's strength lies in its double role as collaborative research tool and social knowledge-sharing site. Users can perform their own research and use Diigo to manage and facilitate those practices, but they can also engage with other users via the built-in social

network and repository of shared bookmarks. In this way, Diigo encourages social knowledge by both taking the individual's needs and desires seriously and providing an online forum for inter-user interaction.

Evernote Corporation. 2017. *Evernote.* **http://evernote.com/.**
Evernote is a platform for capturing and archiving digital content. Applicable content includes formatted text, web pages, images, audio, and handwritten text. In the tool, every individual file or document becomes a note, and these notes can be easily shared, organized, and archived. Although primarily geared toward individual research and project management, Evernote can facilitate collaborative work through sharing practices.

Glass, Geof. 2005. *Marginalia.* **http://webmarginalia.net/.**
Although Marginalia could feasibly be adopted for other endeavours, it was primarily designed with education, collaboration, and online discussion in mind. As a web annotation system, Marginalia integrates with learning management systems like Moodle. Marginalia acts as both a straightforward tool for personal and collaborative annotation as well as a more comprehensive forum discussion. Of note, this tool is open source.

Google Inc. 2012. *Google Drive.* **https://www.google.com/drive/.**
Google Drive is a browser-based application for document storage, creation, and sharing online. More than 30 file types can be saved, and common file types (documents, presentations, spreadsheets) created, in the Google Drive environment. In addition to allowing users to develop and save files online, Google Drive also facilitates easy collaboration, as it enables multiple users to chat, comment, and work on the same document simultaneously. The documents also contain a versioning system for users to revert to previous versions or view specific changes.

Hammond, Adam, and Julian Brooke. 2011–12. *He Do the Police in Different Voices.* **http://hedothepolice.org/.**
Hammond and Brooke created the website *He Do the Police in Different Voices* for the specific exploration of T.S. Eliot's notoriously complex poem, The Waste Land. (The title is an allusion to Eliot's working title for his poem.) So far used only in a classroom setting (at the time of writing), *He Do the Police in Different Voices* encourages students to annotate *The Waste Land* for voice. The website incorporates versions of *The Waste Land* that have already been marked up for voice and automated through an algorithm. Although this website is not a tool, per se, it does demonstrate the various ways in which collaborative

annotation can instigate social knowledge creation; in this case, new insights and explorations are garnered by focusing group work on a shared text.

Harvard University Herbarium and University of Massachusetts Boston (UMASS-Boston) Biodiversity Informatics Lab. 2010–15. *FilteredPush.* **http://wiki.filteredpush.org/wiki/FilteredPush.**
The goal of the FilteredPush project is to create a cross-institutional infrastructure for biologists, particularly taxonomists, making it easier to share and manage digitized natural history collections data. The development of such a network across multiple remote sites would facilitate identification and annotation of specimens (for example, insects or plants), and address issues of quality control and dissemination.

Haystack Group and Massachusetts Institute of Technology (MIT). n.d. *nb.* **http://nb.mit.edu/.**
nb was initially conceived for use in an educational context. It is a web-based annotation tool and service designed with online discussion in mind. nb can be used to write, share, and respond to annotations in PDF files collaboratively. To date, nb has been used primarily in MIT classroom settings.

Massachusetts General Hospital. 2013. *Domeo.* **http://dbmi-icode-01. dbmi.pitt.edu:2020/Domeo/login/auth.**
Domeo encourages social knowledge creation through shared annotation practices. Domeo is an extensible web application for creating and sharing ontology-based annotations on HTML or XML documents. This application facilitates sharing through the Annotation Ontology (AO) RDF framework. Notably, Domeo supports fully automated, semi-automated, and manual annotation, as well as both personal and community annotation with access authorization and control.

*Open Knowledge Foundation. 2009–12. *AnnotateIt / Annotator.* **http:// annotateit.org.**
AnnotateIt is an effective and easy to use system that enables online annotations. A bookmarklet is used to add the JavaScript tool Annotator to any web page; users can then annotate or comment on various elements on the page, and save the annotations to AnnotateIt. This sort of tool readily facilitates social knowledge creation through collaborative annotation. User annotations may contain tags, content created using the Markdown conversion tool, and individual permissions per annotation. Annotator is also easily extensible, allowing for the potential inclusion of more behaviours or features. Of note, the Open Knowledge Foundation has developed many social knowledge

creation tools, including BibServer (https://github.com/okfn/bibserver), CKAN (http://ckan.org/), and TEXTUS (http://textusproject.org/)—all of which are annotated in this bibliography.

_____ 2011–13. *TEXTUS*. http://textusproject.org/.

TEXTUS is an open source platform that aims to encourage online discussion and enhance professional reading environments. More specifically, this service was designed for students, researchers, and teachers to collaboratively work with texts. With TEXTUS, users can individually or collaboratively annotate texts as well as view others' annotations.

Protonotes. 2008. *Protonotes*. http://www.protonotes.com/.

Protonotes is a simple, straightforward collaborative annotation tool for prototype development. Protonotes enables the direct addition of notes onto a prototype, for the purpose of collaborative development. It is free to use and simply requires installing a JavaScript library into the desired prototype. When the installation is complete, anyone who visits the prototype may view, add, edit, or delete notes.

Scholars' Lab (University of Virginia Library). 2012. *Prism*. http://prism. scholarslab.org.

Prism is an open access user-friendly tool for crowdsourcing interpretation where users can highlight different words or sections of the text according to certain predetermined, bounded categories. By allowing the same section of the text to be matched to different categories, Prism demonstrates the multiplicity of possible interpretations while conducting close reading, rather than having one falsely unified category that leaves out space for uncertainty. When annotations are completed, Prism portrays a pie-chart for every word or section, which displays the percentage of all the categories tagged by different users. It can easily be adapted to the classroom environment and is helpful in collectively analyzing literary texts, especially poems with multilayered meanings.

Tejeda, Eddie A. 2008–11. *Digress.it*. http://digress.it/.

Digress.it attempts to alter e-reading practices by facilitating vertical, right-side commenting on online documents. By shifting the comment space from the more conventional blog style (comments below post) to side-by-side text and commentary, Digress.it aims to facilitate greater engagement in online reading environments. In this way, Digress.it strives to emulate the long- standing textual ritual of marginalia. Digress.it is a WordPress plug-in, and thus primarily intended for use on WordPress blogs and sites. Of note,

Digress.it developed from the Institute for the Future of the Book's CommentPress project. The tool is also open source and free to use.

Textensor. 2008. *A.nnotate*. http://a.nnotate.com/index.html.
As a browser-based tool, A.nnotate allows users to privately or publicly annotate and index documents, images, and snapshots of webpages. In this
way, A.nnotate can be used by an individual as a personal indexing tool or by
a group to collaboratively comment on a shared document. A.nnotate facilitates further document management practices, including reviewing drafts,
compiling corrections for revision, and noting passages for future reference.

Whaley, Dan. 2011. *Hypothes.is*. https://hypothes.is/.
Hypothes.is facilitates the annotation of web content and has a reputation
system that ranks the credibility of the annotations. Its goal is to create a
layer of annotations on top of existing knowledge content, and to facilitate
discussions and collaboration. Hypothes.is is an open access platform where
users can contribute and share comments. The primary aim is to create an
open, ubiquitous, interoperable, and lasting tool that will ensure the preservation and reuse of data far into the future.

Zurb. 2011–17. *Bounce*. http://www.bounceapp.com/.
Bounce attempts to improve prototype development via an open, shared
feedback structure. As a ZURBapp, Bounce was created to facilitate productive, collaborative design work. Specifically, Bounce was designed for colleagues to provide each other with feedback on ongoing projects. Users can
upload an image or submit a URL, and comment directly onto this file. In
the framework of collaborative annotation, Bounce could ostensibly be used
to share basic notations on a shared document easily. One may also copy
and paste a Bounce-generated URL for dissemination after commenting on
a page.

2. User-Derived Content

Citizen Cyberscience Centre and Open Knowledge Foundation. 2013.
    ***PyBossa*. http://pybossa.com.**
An open source platform, PyBossa enables the creation of web applications
for individuals to participate in and submit content to. More specifically,
PyBossa is a micro-tasking platform that utilizes crowdsourcing in order
to carry out small, user-derived tasks and contributions. To date, *Crowd-
crafting* (crowdcrafting.org) remains the most notable project developed
on PyBossa.

Gruzd, Anatoliy. 2006–16. *Netlytic.* **http://netlytic.org/.**
Netlytic detects and expresses the innate social networks of online participants based on users' digital tracks. This web-based social network analysis tool summarizes large amounts of text and discovers social networks from electronic communications, including emails, forums, blogs, chats, YouTube, and Twitter. Netlytic allows a user to either capture or import relevant online data, and to analyze this data for emergent themes, trends, and relationships. Furthermore, with Netlytic, users can visualize these communication networks.

*Insemtives. 2009–12. *INSEMTIVES.* **http://insemtives.eu/.**
This suite of tools focuses on the creation of semantic content via incentive-based gaming environments. INSEMTIVES aims to bridge the gap between machine-readable computational data and the necessary limitations of automating semantic content creation tasks. By providing incentives, INSEMTIVES attempts to inspire individuals to manually create, extend, or revise semantic content. This tool is geared toward social knowledge creation through user-generated content and participation.

Jacoby, John James (lead developer). 2009. *BuddyPress.* **http:// buddypress.org/.**
BuddyPress is a social network tool built off its parent project, WordPress. With BuddyPress, a user can instigate a social network customized for various purposes or communities. In this way, BuddyPress actively constructs a framework for social knowledge creation. Of note, BuddyPress is open source, easily extensible, and provides a range of features.

LearningTimes, LLC. 2013. *BadgeOS.* **http://badgeos.org/badgestack/.**
BadgeOS is a free WordPress plug-in that facilitates the creation of rewards- or achievements- based environments. It enables organizations and individuals alike to create sites that incorporate the currently popular practices of structuring digital activities in a game-inspired manner. The Badge Stack add-on helps indicate activities and successes by including rewards and credentials in the form of levels, quests, achievements, and badges. BadgeOS uses the widely recognized credential system from Mozilla Badges. As well, all badges and credentials are shareable through integration with Facebook, Twitter, LinkedIn, blogs, and even individuals' resumes.

Mideast Youth. 2016. *CrowdVoice: Tracking Voices of Protest.* **http:// crowdvoice.org/.**
CrowdVoice is an overtly political web service that harnesses crowdsourcing to track and provide updates on protests around the world. The website

allows protesters to share, and others to view, information, images, video, links, and updates of events. In this way, *CrowdVoice* offers an alternative to standard news outlets and draws attention to corruption, violence, uprisings, and revolutions as they occur. This project is not open source due to risk of persecution for involvement with or contribution to the site. *CrowdVoice* is an exemplary instance of how user- derived content can foster social knowledge creation and even, perhaps, social change.

*Mozilla Foundation. 2011. *Open Badges.* **https://openbadges.org.**
Mozilla's Open Badges is an alternative credential-granting system designed for the public recognition of non-conventional learning and success. Broadly articulated as a democratizing service, Open Badges allows various organizations to accredit their participants within a recognizable system. In an era of Massive Open Online Courses (MOOCs) and citizen scholars, Open Badges embodies the ethos of the decentralized network of contemporary learning, accreditation, and social knowledge creation.

Open Knowledge Foundation. 2016. *CKAN.* **http://ckan.org/.**
Employed by various government catalogues, CKAN is both a web-based data portal and data management system. CKAN supports data publishers (governments, data providers) with services to publish data through a guided process, customize metadata and branding, manage versions, access user analytics, and to store data. As a data portal, CKAN encourages data users (researchers, journalists, programmers, NGO's, citizens) to build extensions, search and tag data sets, engage in a social network, and access metadata and APIs. CKAN's dual role induces social knowledge creation through both user-generated and user- manipulated content. Notably, CKAN is completely open source and easily customizable.

*Roy Rosenzweig Center for History and New Media (George Mason University). 2007–17. *Omeka.* **http://omeka.org.**
Omeka is an example of social knowledge creation through user-driven or generated content. An open source content management system, Omeka was designed to display online digital collections of scholarly editions and cultural heritage artifacts. This content management system acts as a collections management tool and an archival digital collection system, allowing for productive scholarly and non-scholarly exhibitions to develop. Omeka includes an extensive list of features aimed at scholars, museum professionals, librarians, archivists, educators, and other enthusiasts. Of note, the Roy Rosenzweig Center also developed the open bibliography initiative Zotero (included in this annotated bibliography).

Transliteracies Project (University of California Santa Barbara). 2012.
    ***RoSE*. http://rose.english.ucsb.edu/.**
RoSE aims to foster a more networked, holistic environment for humanities research, scholarship, and practices. By combining farmed information from the digital library Project Gutenberg and the semantic knowledge base YAGO with user-generated content, RoSE methodically constructs a social network of collaborators, authors, movements, and works. These relationships are visualized either as a social network graph or in a packed radial style. In this way, users can both contribute to and benefit from the linking of various individuals and texts.

3. Folksonomy Tagging

Association for Computers and the Humanities (ACH) and *ProfHacker*.
    2010. *Digital Humanities Questions & Answers*. http://digitalhumanities.
    org/answers/.
Digital Humanities Questions & Answers (known simply as "DHAnswers") is an online question and answer board for digital humanities practitioners run as a collaborative project by the Association for Computers and the Humanities (ACH) and the *Chronicle of Higher Education* blog, *ProfHacker*. Questions are appropriately tagged as they are asked, thus creating a collection of tags for others to navigate and ideally find answers to their own relevant questions. DHAnswers provides an excellent example of how folksonomy tagging can be harnessed by a specific community in order to foster social knowledge creation on a predetermined subject.

Delicious Media, INC. 2017. *Delicious*. https://del.icio.us/.
Delicious is primarily a social bookmarking site. Users can bookmark various links, websites, or articles on the Internet and share these bookmarks with other Delicious users. Although the default setting is public sharing, users can choose to archive bookmarks privately. Folksonomy develops on Delicious as users tag their selected bookmarks with any desirable metadata terms. Delicious facilitates knowledge creation through a purposefully social environment.

*Huffman, Steve, and Alexis Ohanian. 2005. *Reddit*. https://www.reddit.
    com.
As a popular social news site, Reddit prompts users to tag and submit content. The hierarchy of posts on the front page of the site (as well as the other pages on the site) is decided by a ranking system predicated on both date of submission and voting by other users. Reddit exemplifies social knowledge

creation via folksonomy tagging in a social network environment. Notably, the news site is also open source.

Pinterest, Inc. 2013. *Pinterest*. http://pinterest.com/.
Billed as "the world's catalog of ideas," Pinterest merges folksonomy tagging, inspiration boards, and a classic social network framework. A web-based application, Pinterest encourages sharing through "pinning" or posting image or video collections to a user's pinboard or page. Pins can be freely shared and circulated, multiple users can pin on the same board, and users can follow other users' boards. Notably, boards can be public or private depending on user preferences.

StumbleUpon, Inc. 2017. *StumbleUpon*. http://www.stumbleupon.com.
Founded in 2002, StumbleUpon is a discovery search engine that finds and recommends content based on personal user interests. In this way, users may discover new content based on their already-asserted interests. In order to keep the system running, users are encouraged to rate content while they review it, as peer-sourcing functions determine relevant content. Through collaborative filtering and folksonomy tagging the system organizes and culls user opinions. Notably, StumbleUpon also functions as a social network.

Yahoo Inc. 2005–17. *Flickr*. http://www.flickr.com/.
At the time of writing, Flickr boasted more than 8 billion images and 70 million photographers or active content uploaders on the site. Flickr relies heavily on folksonomy tagging to support its community and induce cross-community media sharing. Users can tag their uploaded photos in order to promote sharing, as well as take advantage of personal indexing capacities by tagging other's images. Notably, institutions such as the White House and NASA also maintain their own Flickr streams.

4. Community Bibliography

Apache Software Foundation. 2010–17. *Apache Subversion*. http:// subversion.apache.org.
Created in 2000, Subversion is an open source, centralized software versioning and revision control system. It differs from its rival version control tool GitHub (annotated below) in that it functions as one repository with many clients; by contrast, each GitHub user has his or her own local repository, and pushes changes to a centralized repository when desired. Like all Apache products, Subversion works and facilitates social knowledge creation following an open source ethos.

DEVONtechnologies, LLC. n.d. *DEVONthink.* **http://www.devontechnologies.com/products/devonthink/overview.html.**
DEVONthink is a proprietary solution created by DEVONtechnologies. It allows users to save and organize documents in one program on their local drive. DEVONthink automatically files and connects related documents, and promotes sharing by enabling users to store their database on a local network or online. Users can also create a bibliographic record for each entry that is then indexed in DEVONthink along with the file.

Drupal. 2006. *Bibliography Module.* **http://drupal.org/project/biblio.**
Bibliography Module, also known as Drupal Scholar, is a Drupal module that enables users to manage and present lists of scholarly publications on Drupal sites using a variety of import and export formats (BibTex, EndNote, MARC, and more). Output is available in most major citation styles and allows for in-line citing of references. Bibliography Module also includes taxonomy integration that allows for higher searchability.

GitHub, Inc. 2017. *GitHub.* **https://github.com.**
As a code repository, GitHub is predicated on transparent and hierarchical project management and organization. GitHub facilitates effective version control by backing up code for a project; allowing collaborative annotation or commenting on lines of code; providing varying levels of access for different team members; hosting unlimited collaborators; and supplying integrated issue tracking. Repositories can be private (secured, limited access) or public (open for community collaboration). GitHub is an exemplary instance of a collaborative project management and indexing tool specifically geared toward digital endeavours.

Grimshaw, Mark. 2003. *WIKINDX.* **http://wikindx.sourceforge.net.**
WIKINDX is a free online bibliography as well as a quotation and note management system. It allows for collaborative use of and contributions to bibliographic data, while also providing features for users to add notes, quotations, and articles. The tool thus functions as reference management software and as a collaborative writing environment. WIKINDX includes search functionalities, allows for attachments to bibliographic resources, exports into most major data and citation styles, and offers customizable plug-ins.

Kaps, Jens-Peter. 2003–9. *Document Database.* **http://docdb.sourceforge. net.**
Document Database is an open source PHP database, written modularly and able to run on users' web servers and with other databases. It can be managed

by multiple administrators and users, and employs BibTeX format. It also allows for various search functions, query types, note sharing, display of user statistics, and uploading and publishing capabilities. Moreover, Document Database includes an agenda management system that associates events and meetings with specific documents.

KDE Group at the University of Kassel, DMIR Group at the University of Würzburg, and L3S Research Center, Hannover. 2006. *BibSonomy.* **http://www.bibsonomy.org.**
BibSonomy is a social bookmarking and publication-sharing system geared to the management of lists of literature. Users can store and organize resources in a public framework, and tag entries with descriptive, user-determined terms. All publications are stored in BibTeX format and can be exported in a variety of ways, including EndNote and HTML.

Massachusetts Institute of Technology (MIT). 2008. *Citeline.* **http://www. simile-widgets.org/wiki/Citeline.**
Citeline allows users to import bibliographies using BibTeX and publish them in the form of an online exhibit. In this way, users may easily create shareable, interactive bibliography exhibits. Users may also select from different background styles for the visual design of their bibliography exhibit.

Mendeley Ltd. 2013. *Mendeley.* **http://dev.mendeley.com.**
Mendeley functions as a free reference management system and an academic social network. Users can generate bibliographies, collaborate with other users, and import resources. The program can be accessed online and as a desktop, iPhone, or iPad application. While the standard tool is free and provides users with 2 GB of web storage space, additional storage can be purchased. The tool also includes a PDF viewer where users can add notes and highlight text. Citations can be exported as BibTeX and into several word processors. The social networking features include newsfeeds, comments, and profile pages. User statistics about papers, authors, and publications may also be viewed.

New Zealand Digital Library Project (University of Waikato). 2005–17. *Greenstone Digital Library Software.* **http://www.greenstone.org.**
Greenstone is an open source software suite for creating and publishing digital library collections online. The software includes command-line tools, as well as a graphical Greenstone Librarian Interface, for users to build collections and assign metadata. User plug-ins enable importing of various digital document formats (including text, html, jpg, MP#, and video).

Open Knowledge Foundation. 2011–12. *BibServer.* **https://github.com/ okfn/bibserver.**
BibServer is open source software that allows for large bibliographic collections managed on tools such as Zotero, Bibsonomy, or Mendeley to be published and shared on the web through a RESTful API and JSON format. The tool allows for collections to be customized and structured using filters. BibServer also offers a variety of visualization options, such as bar charts and bubbles.

Oversity Ltd. 2006. *CiteULike.* **http://www.citeulike.org.**
CiteULike is a free online social bookmarking service for scholarly research. Users can search and discover resources, receive automatic article recommendations, share references, view what others are reading, and store and search a repository of PDFs. CiteULike is structured as a folksonomy, allowing users to tag references and thus organize their libraries. In addition to adding tags, users can comment on and rate resources. Citation information can be automatically imported from a number of popular databases, such as JSTOR and arXiv.org, and citations can also be imported or adjusted manually, or transferred to other reference management systems (e.g., EndNote or Zotero).

Roy Rosenzweig Center for History and New Media (George Mason University). 2006–17. *Zotero.* **https://www.zotero.org/.**
Zotero is an open source reference management system for users to store citations and other content in a variety of file formats. Most library catalogues and common online research environments contain Zotero links, and Zotero integrates with word processors and other writing environments (e.g., email and Google Drive), making it easy to save reference information while working. Users can also assign tags to library items and organize research into collections and subcollections. The tool functions and automatically synchronizes across multiple devices and web browsers. One of the capabilities that differentiates Zotero is the ability to create topical research groups that can house shared libraries, notes, and discussions, offering a collaborative research environment.

Slashdot Media. 2017. *SourceForge.* **http://sourceforge.net.**
SourceForge is a web-based source code repository comprising a suite of tools dedicated to facilitating open source software development and dissemination. SourceForge resources include version control, integrated issue tracking, threaded discussion forums, documentation, download statistics, a code repository, and an open source directory. SourceForge induces social

knowledge creation by hosting and indexing open source projects and providing easeful access to these projects for the community at large. Notably, SourceForge was the first service to offer free hosting for open source projects.

University of Southampton. 2017. *EPrints*. http://www.eprints.org.
EPrints is open-source software for creating open access repositories. It is a command-line application written in Perl. The database repository can be controlled using HTML, CSS, and inline images. EPrints allows for metadata harvesting and is most commonly used for institutional repositories and scientific journals. The software allows for data importing and exporting, object conversion for search engine indexing, and various user interface widgets.

5. Shared Text Analysis

Baron, Alistair. 2008–17. *VARD 2*. http://ucrel.lancs.ac.uk/vard/about/.
VARD 2 is interactive software that permits users to identify and replace spelling variations in historical texts, primarily Early Modern English texts. Spelling variations can be adjusted manually, replaced automatically, or defined semi-automatically by manually training the tool.

Harvard University Library Lab. 2011. *Highbrow*. https://osc.hul.harvard. edu/liblab/projects/highbrow-textual-annotation-browser.
Highbrow is a textual annotation browser and visualization tool. It visualizes the density of scholarly annotations and references in individual texts, and can compare multiple texts to indicate patterns or highlight areas of interest for scholars. Users can view the visualizations at a higher level of quality that indicates density, or else zoom in for more detailed information. Highbrow functions for textual annotations as well as video and audio annotations.

Northwestern University. 2004–13. *WordHoard*. http://wordhoard. northwestern.edu/userman/index.html.
WordHoard is a free Java application developed by Northwestern University that enables tagging and annotations of large texts or transcribed speech. Currently, WordHoard is aimed toward early Greek epics and early modern English plays, but also includes texts by authors such as Chaucer. WordHoard allows users to easily annotate and analyze texts by looking at word frequency, lemmatization, and text comparison, or by applying custom queries.

Sinclair, Stéfan, Geoffrey Rockwell, and the Voyant Tools Team. 2012. *Voyant Tools*. http://voyant-tools.org.
Voyant is an online text analysis environment. Users can submit texts in many formats from various locations (e.g., by using URLs to indicate entire

web pages). Voyant analyzes single or multiple texts and displays word usage by indicating frequency of words, visualizing usage of words, and showing placements of words throughout documents.

TAPoR Team. 2015. *TAPoR* (version 3.0). http://www.tapor.ca.
TAPoR (Text Analysis Portal for Research) is a collection of textual studies tools for scholars and researchers. The site functions as a portal to a number of tools relevant to textual studies scholars. Each tool listed is tagged with keywords, includes a short description, details information about documentation and tool attributes, and displays user ratings and comments.

University of Hamburg. 2009. *CATMA*. http://catma.de.
CATMA is a web-based text analysis and literary research application that permits scholars to work collaboratively by exchanging analytical results online. The application boasts a number of features: users can apply analytical categories and tags; search the text using Query Builder; set predefined statistical and non-statistical analytical functions; visualize text attributes and findings; and share documents, tagsets, and markup. CATMA consists of three modules: the Tagger for markup and tagging of a text, the Analyzer for queries and a variety of text analysis functions, and the Visualizer to create charts and other visualizations of analysis results.

Vision Critical Communications. 2013. *DiscoverText*. http://discovertext. com.
DiscoverText is a proprietary software solution that enables cloud-based, collaborative text analysis. This tool is primarily employed by the public and private sectors to analyze and gather insights about user, consumer, and employee activity and engagement. The software merges data from numerous sources, including text files, email, surveys, and online platforms (e.g., Facebook, Twitter, Google+, and blogs).

6. A Complete Alphabetical List of Selections

Abilian SAS. 2010–15. *co-ment*. http://www.co-ment.com/.

Apache Software Foundation. 2010–17. *Apache Subversion*. http://subversion. apache.org.

Association for Computers and the Humanities (ACH) and *ProfHacker*. 2010. *Digital Humanities Questions & Answers*. http://digitalhumanities.org/ answers/.

Baron, Alistair. 2008–17. *VARD 2*. http://ucrel.lancs.ac.uk/vard/about/.

Citizen Cyberscience Centre and Open Knowledge Foundation. 2013. *PyBossa.* http://pybossa.com.

Delicious Media, INC. 2017. *Delicious.* https://del.icio.us/.

DEVONtechnologies, LLC. n.d. *DEVONthink.* http://www.devontechnologies.com/products/devonthink/overview.html.

Diigo, Inc. *Diigo.* 2017. https://www.diigo.com/.

Drupal. 2006. *Bibliography Module.* http://drupal.org/project/biblio.

Evernote Corporation. 2017. *Evernote.* http://evernote.com/.

GitHub, Inc. 2017. *GitHub.* https://github.com.

Glass, Geof. 2005. *Marginalia.* http://webmarginalia.net/.

Google Inc. 2012. *Google Drive.* https://www.google.com/drive/.

Grimshaw, Mark. 2003. *WIKINDX.* http://wikindx.sourceforge.net.

Gruzd, Anatoliy. 2006–16. *Netlytic.* http://netlytic.org/.

Hammond, Adam, and Julian Brooke. 2011–12. *He Do the Police in Different Voices.* http://hedothepolice.org/.

Harvard University Herbarium and University of Massachusetts Boston (UMASS-Boston) Biodiversity Informatics Lab. 2010–15. *FilteredPush.* http://wiki.filteredpush.org/wiki/FilteredPush.

Harvard University Library Lab. 2011. *Highbrow.* https://osc.hul.harvard.edu/liblab/projects/highbrow-textual-annotation-browser.

Haystack Group and Massachusetts Institute of Technology (MIT). n.d. *nb.* http://nb.mit.edu/.

Huffman, Steve, and Alexis Ohanian. 2005. *Reddit.* https://www.reddit.com.

Insemtives. 2009–12. *INSEMTIVES.* http://insemtives.eu/.

Jacoby, John James (lead developer). 2009. *BuddyPress.* http://buddypress.org/.

Kaps, Jens-Peter. 2003–9. *Document Database.* http://docdb.sourceforge.net.

KDE Group at the University of Kassel, DMIR Group at the University of Würzburg, and L3S Research Center, Hannover. 2006. *BibSonomy*. http://www.bibsonomy.org.

LearningTimes, LLC. 2013. *BadgeOS*. http://badgeos.org/badgestack/.

Massachusetts General Hospital. 2013. *Domeo*. http://dbmi-icode-01.dbmi.pitt.edu:2020/Domeo/login/auth.

Massachusetts Institute of Technology (MIT). 2008. *Citeline*. http://www.simile-widgets.org/wiki/Citeline.

Mendeley Ltd. 2013. *Mendeley*. http://dev.mendeley.com.

Mideast Youth. 2016. *CrowdVoice: Tracking Voices of Protest*. http://crowdvoice.org/.

Mozilla Foundation. 2011. *Open Badges*. https://openbadges.org.

New Zealand Digital Library Project (University of Waikato). 2005–17. *Greenstone Digital Library Software*. http://www.greenstone.org.

Northwestern University. 2004–13. *WordHoard*. http://wordhoard.northwestern.edu/userman/index.html.

Open Knowledge Foundation. 2009–12. *AnnotateIt / Annotator*. http://annotateit.org.

_____ 2011–12. *BibServer*. https://github.com/okfn/bibserver.

_____ 2011–13. *TEXTUS*. http://textusproject.org/.

_____ 2016. *CKAN*. http://ckan.org/.

Oversity Ltd. 2006. *CiteULike*. http://www.citeulike.org.

Pinterest, Inc. 2013. *Pinterest*. http://pinterest.com/.

Protonotes. 2008. *Protonotes*. http://www.protonotes.com/.

Roy Rosenzweig Center for History and New Media (George Mason University). 2006–17. *Zotero*. https://www.zotero.org/.

_____ 2007–17. *Omeka*. http://omeka.org.

Scholars' Lab (University of Virginia Library). 2012. *Prism.* http://prism. scholarslab.org.

Sinclair, Stéfan, Geoffrey Rockwell, and the Voyant Tools Team. 2012. *Voyant Tools.* http://voyant-tools.org.

Slashdot Media. 2017. *SourceForge.* http://sourceforge.net.

StumbleUpon, Inc. 2017. *StumbleUpon.* http://www.stumbleupon.com.

TAPoR Team. 2015. *TAPoR* (version 3.0). http://www.tapor.ca.

Tejeda, Eddie A. 2008–11. *Digress.it.* http://digress.it/.

Textensor. 2008. *A.nnotate.* http://a.nnotate.com/index.html.

Transliteracies Project (University of California Santa Barbara). 2012. *RoSE.* http://rose.english.ucsb.edu/.

University of Hamburg. 2009. *CATMA.* http://catma.de.

University of Southampton. 2017. *EPrints.* http://www.eprints.org.

Vision Critical Communications. 2013. *DiscoverText.* http://discovertext.com.

Whaley, Dan. 2011. *Hypothes.is.* https://hypothes.is/.

Yahoo Inc. 2005–17. *Flickr.* http://www.flickr.com/.

Zurb. 2011–17. *Bounce.* http://www.bounceapp.com/.

Complete Alphabetical List of Bibliography

Aarseth, Espen J. 1997. "Introduction." In *Cybertext: Perspectives on Ergodic Literature*, 1–23. Baltimore: Johns Hopkins University Press.

_____ 2012. "A Narrative Theory of Games." In *Proceedings of the International Conference on the Foundations of Digital Games (FDG '12)*, 129–33. New York: ACM. doi:10.1145/2282338.2282365.

Abilian SAS. 2010–15. *co-ment.* http://www.co-ment.com/.

Althusser, Louis. 1971. "Ideology and Ideological State Apparatuses (Notes Towards an Investigation)." In *Lenin and Philosophy and Other Essays*, translated by Ben Brewster, 127–86. New York: Monthly Review Press.

Ancient World Mapping Center, Stoa Consortium, and Institute for the Study of the Ancient World. 2000. *Pleiades.* http://pleiades.stoa.org/.

Andersen, Christian Ulrik, and Søren Bro Pold. 2014. "Post-digital Books and Disruptive Literary Machines." *Formules / Revue des Créations Formelles et Littératures à Contraintes* 18: 169–88.

Ang, Ien. 2004. "Who Needs Cultural Research?" In *Cultural Studies and Practical Politics: Theory, Coalition Building, and Social Activism*, edited by Pepi Leystina, 477 83. New York: Blackwell.

Anthropy, Anna. 2012. *Rise of the Videogame Zinesters: How Freaks, Normals, Amateurs, Artists, Dreamers, Dropouts, Queers, Housewives, and People Like You Are Taking Back an Art Form.* New York: Seven Stories Press.

Apache Software Foundation. 2010–17. *Apache Subversion.* http://subversion.apache.org.

Arbuckle, Alyssa, and Alex Christie, with the ETCL, INKE, and MVP Research Groups. 2015. "Intersections Between Social Knowledge Creation and Critical Making." *Scholarly and Research Communication* 6 (3): n.p. http://src-online.ca/index.php/src/article/view/200.

Arbuckle, Alyssa, Constance Crompton, and Aaron Mauro. 2014. Introduction: "Building Partnerships to Transform Scholarly Publishing." *Scholarly and*

Research Communication 5 (4): n.p. http://src-online.ca/index.php/src/article/view/195.

Arbuckle, Alyssa, Aaron Mauro, and Lynne Siemens. 2015. Introduction: "From Technical Standards to Research Communities: Implementing New Knowledge Environments Gatherings, Sydney 2014 and Whistler 2015." *Scholarly and Research Communication* 6 (2): n.p. http://src-online.ca/index.php/src/article/view/232.

Association for Computers and the Humanities (ACH) and *ProfHacker*. 2010. *Digital Humanities Questions & Answers.* http://digitalhumanities.org/answers/.

Avila, Maria, with contributions from Alan Knoerr, Nik Orlando, and Celestina Castillo. 2010. "Community Organizing Practices in Academia: A Model, and Stories of Partnerships." *Journal of Higher Education Outreach and Engagement* 14 (2): 37–63. http://openjournals.libs.uga.edu/index.php/jheoe/article/view/43/38.

Bachelard, Gaston. 1969. *The Poetics of Space.* Translated by Maria Jolas. Boston: Beacon Press.

Bailey, Moya Z. 2011. "All the Digital Humanists Are White, All the Nerds Are Men, But Some of Us Are Brave." *Journal of Digital Humanities* 1 (1): n.p. http://journalofdigitalhumanities.org/1-1/all-the-digital-humanists-are-white-all-the-nerds-are-men-but-some-of-us-are-brave-by-moya-z-bailey/.

Ball, John Clement. 2010. "Definite Article: Graduate Student Publishing, Pedagogy, and the Journal as Training Ground." *Canadian Literature* 204: 160–62.

Balsamo, Anne. 2011. Introduction: "Taking Culture Seriously in the Age of Innovation." In *Designing Culture: The Technological Imagination at Work,* 2–25. Durham, NC: Duke University Press.

Baron, Alistair. 2008–17. *VARD 2.* http://ucrel.lancs.ac.uk/vard/about/.

Bath, Jon, and Scott Schofield. 2015. "The Digital Book." In *The Cambridge Companion to the History of the Book,* edited by Leslie Howsam, 181–95. Cambridge: Cambridge University Press.

Bazerman, Charles. 1991. "How Natural Philosophers Can Cooperate: The Literary Technology of Coordinated Investigation in Joseph Priestley's

History and Present State of Electricity (1767)." In *Textual Dynamics of the Professions: Historical and Contemporary Studies of Writing in Professional Communities*, edited by Charles Bazerman and James Paradis, 13–44. Madison: University of Wisconsin Press.

Beller, Jonathan. 2006. *The Cinematic Mode of Production: Attention Economy and the Society of the Spectacle*. Hanover, NH: Dartmouth College Press.

_____ 2006/07. "Paying Attention." *Cabinet* 24: n.p. http://www. cabinetmagazine.org/issues/24/beller.php.

Benkler, Yochai. 2003. "Freedom in the Commons: Towards a Political Economy of Information." *Duke Law Journal* 52 (6): 1245–76. http:// scholarship.law.duke.edu/dlj/vol52/iss6/3.

Berkenkotter, Carol, Thomas N. Huckin, and John Ackerman. 1991. "Social Context and Socially Constructed Texts: The Initiation of a Graduate Student into a Writing Research Community." In *Textual Dynamics of the Professions: Historical and Contemporary Studies of Writing in Professional Communities*, edited by Charles Bazerman and James Paradis, 191–215. Madison: University of Wisconsin Press.

Berry, David M. 2011. "The Computational Turn: Thinking About the Digital Humanities." *Culture Machine* 12: n.p. http://www.culturemachine.net/ index.php/cm/article/view/440/470.

_____ 2012. "The Social Epistemologies of Software." *Social Epistemology: A Journal of Knowledge, Culture and Policy* 26 (3–4): 379–98. doi:10.1080/02 691728.2012.727191.

Besser, Howard. 2004. "The Past, Present, and Future of Digital Libraries." In *A Companion to Digital Humanities*, edited by Susan Schreibman, Raymond G. Siemens, and John Unsworth, 557–75. Oxford: Blackwell.

Biagioli, Mario. 2002. "From Book Censorship to Academic Peer Review." *Emergences* 12 (1): 11–45. doi:10.1080/1045722022000003435.

Bijker, Wiebe E., and John Law. 1992. "General Introduction." In *Shaping Technology/Building Society: Studies in Sociotechnical Change*, edited by Wiebe E. Bijker and John Law, 1–14. Cambridge, MA: MIT Press.

Bjork, Staffan, and Jussi Holopainen. 2005. *Patterns in Game Design*. Hingham, MA: Charles River Media.

Blizzard Entertainment, Inc. 2005. *World of Warcraft* (WoW) [video game]. https://worldofwarcraft.com/en-us/.

Bogost, Ian. 2007. *Persuasive Games: The Expressive Power of Videogames.* Cambridge, MA: MIT Press.

_____ 2011a. *How To Do Things With Videogames.* Minneapolis: University of Minnesota Press.

_____ 2011b. "Persuasive Games: Exploitationware" [blog post]. *Gamasutra.* http://www.gamasutra.com/view/feature/6366/persuasive_games_exploitationware.php.

_____ 2012. *Alien Phenomenology, or, What It's Like to Be a Thing.* Minneapolis: University of Minnesota Press.

Bolter, Jay David. 2007. "Digital Media and Art: Always Already Complicit?" *Criticism* 49 (1): 107–19. doi:10.1353/crt.2008.0013.

Boot, Peter. 2012. "Literary Evaluation in Online Communities of Writers and Readers." *Scholarly and Research Communication* 3 (2): n.p. http://src-online.ca/index.php/src/article/view/77/90.

Borgman, Christine L. 2007. *Scholarship in the Digital Age: Information, Infrastructure, and the Internet.* Cambridge, MA: MIT Press.

Bourdieu, Pierre. 1993. "The Field of Cultural Production, or: The Economic World Reversed." In *The Field of Cultural Production: Essays on Art and Literature,* edited and translated by Randal Johnson, 29–73. New York: Columbia University Press.

Bowen, William R., Constance Crompton, and Matthew Hiebert. 2014. "Iter Community: Prototyping an Environment for Social Knowledge Creation and Communication." *Scholarly and Research Communication* 5 (4): n.p. http://src-online.ca/index.php/src/article/view/193/360.

Brant, Claire. 2011. "The Progress of Knowledge in the Regions of Air?: Divisions and Disciplines in Early Ballooning." *Eighteenth-Century Studies* 45 (1): 71–86. doi:10.1353/ecs.2011.0050.

Brooks, Kevin. 2002. "National Culture and the First-Year English Curriculum: A Historical Study of 'Composition' in Canadian Universities." *American Review of Canadian Studies* 32 (4): 673–94. doi:10.1080/02722010209481679.

Brown, David W. 1995. "The Public/Academic Disconnect." In *Higher Education Exchange Annual*, 38–42. Dayton, OH: Kettering Foundation.

Buehl, Jonathan, Tamar Chute, and Anne Fields. 2012. "Training in the Archives: Archival Research as Professional Development." *College Composition and Communication* 64 (2): 274–305.

Burdick, Anne, Johanna Drucker, Peter Lunenfeld, Todd Presner, and Jeffrey Schnapp. 2012. "The Social Life of the Digital Humanities." In *Digital_Humanities*, 73–98. Cambridge, MA: MIT Press.

Burke, Peter. 2000. *A Social History of Knowledge: From Gutenberg to Diderot.* Cambridge: Polity Press.

_____ 2012. *A Social History of Knowledge II: From the Encyclopédie to Wikipedia.* Cambridge: Polity Press.

Caillois, Roger. (1961) 2001. *Man, Play, and Games.* Translated by Meyer Barash. New York: Free Press of Glencoe. Reprint, Chicago: University of Illinois Press.

Cao, Qilin, Yong Lu, Dayong Dong, Zongming Tang, and Yongqiang Li. 2013. "The Roles of Bridging and Bonding in Social Media Communities." *Journal of the American Society for Information Science and Technology* 64 (8): 1671–81. doi:10.1002/asi.22866.

Carletti, Laura, Derek McAuley, Dominic Price, Gabriella Giannachi, and Steve Benford. 2013. "Digital Humanities and Crowdsourcing: An Exploration." Museums and the Web 2013 Conference. Portland: *Museums and the Web LLC.* http://mw2013.museumsandtheweb.com/paper/digital-humanities-and-crowdsourcing-an-exploration-4/.

Carlton, Susan Brown. 1995. "Composition as a Postdisciplinary Formation." *Rhetoric Review* 14 (1): 78–87. doi:10.1080/07350199509389053.

Carson, Stephen, and Jan Philipp Schmidt. 2012. "The Massive Open Online Professor." *Academic Matters: The Journal of Higher Education*, n.p. http://www.academicmatters.ca/2012/05/the-massive-open-online-professor/.

Causer, Tim, and Melissa Terras. 2014. "Crowdsourcing Bentham: Beyond the Traditional Boundaries of Academic History." *International Journal of Humanities and Arts Computing* 8 (1): 46–64. doi:10.3366/ijhac.2014.0119.

Causer, Tim, Justin Tonra, and Valerie Wallace. 2012. "Transcription Maximized; Expense Minimized? Crowdsourcing and Editing *The Collected Works of Jeremy Bentham*." *Digital Scholarship in the Humanities* (formerly *Literary and Linguistic Computing*) 27 (2): 119–37. doi:10.1093/llc/fqs004.

Causer, Tim, and Valerie Wallace. 2012. "Building a Volunteer Community: Results and Findings from *Transcribe Bentham*." *Digital Humanities Quarterly* 6 (2): n.p. http://digitalhumanities.org:8081/dhq/vol/6/2/000125/000125.html.

CCP Games. 2003. *EVE Online* [video game]. http://www.eveonline.com/.

Center for Interdisciplinary Inquiry and Innovation in Sexual and Reproductive Health (Ci3), University of Chicago. 2013. *The Source* [video game]. https://ci3.uchicago.edu/portfolio/the-source/.

Chamberlin, Barbara, Jesús Trespalacios, and Rachel Gallagher. 2014. "Bridging Research and Game Development: A Learning Games Design Model for Multi-Game Projects." In *Educational Technology Use and Design for Improved Learning Opportunities*, edited by Mehdi Khosrow-Pour, 151–71. Hershey, PA: IGI Global. doi:10.4018/978-1-4666-6102-8.ch008.

Chang, Edmond. 2012. "Video+Game+Other+Media: Video Games and Remediation" [blog post]. *Critical Gaming Project.* https://depts.washington.edu/critgame/wordpress/2012/01/videogameothermedia-video-games-and-remediation/.

Chapman, Owen, and Kim Sawchuk. 2015. "Creation-as-Research: Critical Making in Complex Environments." *RACAR: Revue d'art canadienne / Canadian Art Review* 40 (1): 49–52. http://www.jstor.org/stable/24327426.

Chun, Wendy Hui Kyong. 2004. "On Software, or the Persistence of Visual Knowledge." *Grey Room* 18: 26–51. doi:10.1162/1526381043320741.

Christie, Alex, and the INKE and MVP Research Groups. 2014. "Interdisciplinary, Interactive, and Online: Building Open Communication Through Multimodal Scholarly Articles and Monographs." *Scholarly and Research Communication* 5 (4): n.p. http://src-online.ca/index.php/src/article/view/190.

Citizen Cyberscience Centre and Open Knowledge Foundation. 2013. *PyBossa.* http://pybossa.com.

Clement, Tanya. 2011. "Knowledge Representation and Digital Scholarly Editions in Theory and Practice." *Journal of the Text Encoding Initiative* 1: n.p. doi:10.4000/jtei.203.

Cohen, Daniel J. 2008. "Creating Scholarly Tools and Resources for the Digital Ecosystem: Building Connections in the Zotero Project." *First Monday* 13 (8): n.p. doi:10.5210/fm.v13i8.2233.

_____ 2012. "The Social Contract of Scholarly Publishing." In *Debates in the Digital Humanities*, edited by Matthew K. Gold, 319–21. Minneapolis: University of Minnesota Press. http://dhdebates.gc.cuny.edu/debates/text/27.

Cohen, Daniel J., and Tom Scheinfeldt. 2013. "Preface." In *Hacking the Academy: New Approaches to Scholarship and Teaching from Digital Humanities*, edited by Daniel J. Cohen and Tom Scheinfeldt, 3–5. Ann Arbor: University of Michigan Press. doi:10.3998/dh.12172434.0001.001.

Coleman, Susan L., Ellen S. Menaker, Jennifer McNamara, and Tristan E. Johnson. 2015. "Communication for Stronger Learning Game Design." In *Design and Development of Training Games*, edited by Talib S. Hussain and Susan L. Coleman. New York: Cambridge University Press, 31–54. doi:10.1017/CBO9781107280137.003.

Crompton, Constance, Alyssa Arbuckle, Raymond G. Siemens, and the *Devonshire MS* Editorial Group. 2013. "Understanding the Social Edition Through Iterative Implementation: The Case of the Devonshire MS (BL Add MS 17492)." *Scholarly and Research Communication* 4 (3): n.p. http://src-online.ca/index.php/src/article/view/118/311.

Crompton, Constance, Raymond G. Siemens, and Alyssa Arbuckle, with the INKE Research Group. 2015. "Enlisting 'Vertues Noble & Excelent': Behavior, Credit, and Knowledge Organization in the Social Edition." *Digital Humanities Quarterly* 9 (2): n.p. http://www.digitalhumanities.org/dhq/vol/9/2/000202/000202.html.

Crowley, Dennis, and Naveen Selvadurai. 2009. *Foursquare*. New York: Foursquare. https://foursquare.com.

Cyber Creations Inc. 2002. *MMORPG.com* [video game website]. http://www.mmorpg.com/.

Danforth, Liz. 2011. "Gamification and Libraries." *Library Journal* 136 (3): 84–85.

Davidson, Cathy N. 2011. "Why Badges? Why Not?" [blog post]. *HASTAC.* https://www.hastac.org/blogs/cathy-davidson/2011/09/16/why-badges-why-not.

Davidson, Cathy N., and David Theo Goldberg. 2004. "Engaging the Humanities." *Profession:* 42–62. doi:10.1632/074069504X26386.

De Carvalho, Carlos Rosemberg Maia, and Elizabeth S. Furtado. 2012. "Wikimarks: An Approach Proposition for Generating Collaborative, Structured Content from Social Networking Sharing on the Web." In *Proceedings of the 11th Brazilian Symposium on Human Factors in Computing Systems (IHC '12),* 95–98. Porto Alegre, Brazil: Brazilian Computer Society.

Delicious Media, INC. 2017. *Delicious.* https://del.icio.us/.

De Paoli, Stefano, Nicolò De Uffici, and Vincenzo D'Andrea. 2012. "Designing Badges for a Civic Media Platform: Reputation and Named Levels." In *Proceedings of the 26th Annual BCS Interaction Specialist Group Conference on People and Computers (BCS-HCI '12).* Swinton, UK: British Computer Society, 59–68.

De Roure, David. 2014. "The Future of Scholarly Communications." *Insights* 27 (3): 233–38. doi:10.1629/2048-7754.171.

Deterding, Sebastian. 2012. "Gamification: Designing for Motivation." *Interactions* 19 (4): 14–17. doi:10.1145/2212877.2212883.

_____ 2015. "The Lens of Intrinsic Skill Atoms: A Method for Gameful Design." *Human-Computer Interaction* 30 (3–4): 294–335. doi:10.1080/0737 0024.2014.993471.

Deterding, Sebastian, Dan Dixon, Rilla Khalad, and Lennart E. Nacke. 2011. "From Game Design Elements to Gamefulness: Defining 'Gamification.'" In *Proceedings of the 15th International Academic MindTrek Conference: Envisioning Future Media Environments (MindTrek '11),* 9–15. New York: ACM. doi:10.1145/2181037.2181040.

DEVONtechnologies, LLC. n.d. *DEVONthink.* http://www.devontechnologies.com/products/devonthink/overview.html.

Dickey, Michele D. 2007. "Game Design and Learning: A Conjectural Analysis of How Massively Multiple Online Role-Playing Games (MMORPGS) Foster Intrinsic Motivation." *Educational Technology Research and Development* 55 (3): 253–73.

Diigo, Inc. *Diigo.* 2017. https://www.diigo.com/.

Douma, Michael. 2011. "What is Gamification?" [blog post]. *Idea.* http://www.idea.org/blog/2011/10/20/what-is-gamification/.

Drucker, Johanna. 2003. "Designing Ivanhoe." *TEXT Technology* 12 (2): 19–41. http://texttechnology.mcmaster.ca/pdf/vol12_2_03.pdf.

_____ 2006. "Graphical Readings and the Visual Aesthetics of Textuality." *TEXT: An Interdisciplinary Annual of Textual Studies* 16: 267–76. http://www.jstor.org/stable/30227973.

_____ 2009. "From Digital Humanities to Speculative Computing." In *SpecLab: Digital Aesthetics and Projects in Speculative Computing,* 3–18. Chicago: University of Chicago Press.

_____ 2011a. "Humanities Approaches to Graphical Display." *Digital Humanities Quarterly* 5 (1): n.p. http://www.digitalhumanities.org/dhq/vol/5/1/000091/000091.html.

_____ 2011b. "Humanities Approaches to Interface Theory." *Culture Machine* 12: 1–20. http://www.culturemachine.net/index.php/cm/article/viewArticle/434.

_____ 2012. "Humanistic Theory and Digital Scholarship." In *Debates in the Digital Humanities,* edited by Matthew K. Gold, 85–95. Minneapolis: University of Minnesota Press. http://dhdebates.gc.cuny.edu/debates/text/34.

Drucker, Johanna, and Jerome McGann. 2000. *Ivanhoe.* SpecLab (University of Virginia). http://www.ivanhoegame.org/?page_id=21.

Drupal. 2006. *Bibliography Module.* http://drupal.org/project/biblio.

Dyer-Witheford, Nick, and Greig de Peuter. 2009. *Games of Empire: Global Capitalism and Video Games.* Minneapolis: University of Minnesota Press.

Eagleton, Terry. 2010. "The Rise of English." In *The Norton Anthology of Theory and Criticism*, edited by Vincent B. Leitch, 2140–46. New York: W.W. Norton.

Edwards, Charlie. 2012. "The Digital Humanities and Its Users." In *Debates in the Digital Humanities*, edited by Matthew K. Gold, 213–32. Minnesota: University of Minnesota Press. http://dhdebates.gc.cuny.edu/debates/text/31.

Eisenstein, Elizabeth L. 1979. *The Printing Press as an Agent of Change: Communications and Cultural Transformations in Early Modern Europe.* Cambridge: Cambridge University Press.

Ellison, Julie. 2008. "The Humanities and the Public Soul." *Antipode* 40 (3): 463–71. doi:10.1111/j.1467-8330.2008.00615.x.

Ellison, Julie, and Timothy Eatman. 2008. *Scholarship in Public: Knowledge Creation and Tenure Policy in the Engaged University.* Syracuse, NY: Imagining America. http://imaginingamerica.org/wp-content/uploads/2011/05/TTI_FINAL.pdf.

Evernote Corporation. 2017. *Evernote.* http://evernote.com/.

Farland, Maria M. 1996. "Academic Professionalism and the New Public Mindedness." *Higher Education Exchange* Annual: 51–57. http://www.unz.org/Pub/HigherEdExchange-1996q1-00051.

Fernheimer, Janice W., Lisa Litterio, and James Hendler. 2011. "Transdisciplinary ITexts and the Future of Web-scale Collaboration." *Journal of Business and Technical Communication* 25 (3): 322–37. doi:10.1177/1050651911400710.

Ferrara, John. 2012. *Playful Design: Creating Game Experiences in Everyday Interfaces.* Brooklyn, NY: Rosenfeld Media.

Fisher, Caitlin. 2015. "Mentoring Research-Creation: Secrets, Strategies, and Beautiful Failures." *RACAR: Revue d'art canadienne / Canadian Art Review* 40 (1): 46–49. http://www.jstor.org/stable/24327425.

Fitzpatrick, Kathleen. 2007. "CommentPress: New (Social) Structures for New (Networked) Texts." *Journal of Electronic Publishing* 10 (3): n.p. doi:10.3998/3336451.0010.305.

_____ 2009. "Peer-To-Peer Review and the Future of Scholarly Authority." *Cinema Journal* 48 (2): 124–29. doi:10.1353/cj.0.0095.

_____ 2011. *Planned Obsolescence: Publishing, Technology, and the Future of the Academy*. New York: New York University Press. ("Introduction: Obsolescence," 1–14, and Chapter 3: "Texts," 89–120, are accessible at http://raley.english.ucsb.edu/wp-content2/uploads/234/Fitzpatrick. pdf.)

_____ 2012a. "Beyond Metrics: Community Authorization and Open Peer Review." In *Debates in the Digital Humanities*, edited by Matthew K. Gold, 452–59. Minneapolis: University of Minnesota Press. http:// dhdebates.gc.cuny.edu/debates/text/7.

_____ 2012b. "The Humanities, Done Digitally." In *Debates in the Digital Humanities*, edited by Matthew K. Gold, 12–15. Minneapolis: University of Minnesota Press. http://dhdebates.gc.cuny.edu/debates/text/30.

Fjällbrant, Nancy. 1997. "Scholarly Communication—Historical Development and New Possibilities." In *Proceedings of the IATUL Conferences*. West Lafayette, IN: Purdue University Libraries e-Pubs. http://docs.lib. purdue.edu/cgi/viewcontent.cgi?article=1389&context=iatul.

Flanders, Julia. 2005. "Detailism, Digital Texts, and the Problem of Pedantry." *TEXT Technology* 14 (2): 41–70. http://texttechnology.mcmaster.ca/pdf/ vol14_2/flanders14-2.pdf.

_____ 2009. "The Productive Unease of 21st-Century Digital Scholarship." *Digital Humanities Quarterly* 3 (3): n.p. http://www. digitalhumanities.org/dhq/vol/3/3/000055/000055.html.

_____ 2012. "Time, Labor, and 'Alternate Careers' in Digital Humanities Knowledge Work." In *Debates in the Digital Humanities*, edited by Matthew K. Gold, 292–308. Minneapolis: University of Minnesota Press. http:// dhdebates.gc.cuny.edu/debates/text/26.

Foucault, Michel. 1977. *Discipline and Punish: The Birth of the Prison*. Translated by Alan Sheridan. London: Allen Lane and Penguin Books.

Franklin, Michael J., Donald Kossman, Tim Kraska, Sukriti Ramesh, and Reynold Xin. 2011. "CrowdDB: Answering Queries with Crowdsourcing." In *Proceedings of the 2011 ACM SIGMOD International Conference on Management of Data (SIGMOD/PODS '11)*, 61–72. New York: ACM.

Fraser, Nancy. 1990. "Rethinking the Public Sphere: A Contribution to the Critique of Actually Existing Democracy." *Social Text* (25, 26): 56–80. http://www.jstor.org/stable/466240.

Freeman, Jo. 1972. "The Tyranny of Structurelessness." *The Second Wave* 2 (1): n.p. http://www.jofreeman.com/joreen/tyranny.htm.

Galey, Alan, and Stan Ruecker. 2010. "How a Prototype Argues." *Digital Scholarship in the Humanities* (formerly *Literary and Linguistic Computing*) 25 (4): 405–24. doi:10.1093/llc/fqq021.

Galloway, Alexander R. 2006. *Gaming: Essays on Algorithmic Culture.* Minneapolis: University of Minnesota Press.

Galloway, Alexander R., Carolyn Kane, Adam Parrish, Daniel Perlin, DJ / rupture and Matt Shadetek, Mushon Zer-Aviv, and the RSG Collective. *Kriegspiel* [video game]. New York University. http://r-s-g.org/kriegspiel/index.php.

Gamification Wiki. "Gamification." https://badgeville.com/wiki/gamification. Dublin, CA: Badgeville.

Garson, Marjorie. 2008. "ACUTE: The First Twenty-Five Years, 1957–1982." *English Studies in Canada* 34 (4): 21–43. https://ejournals.library.ualberta.ca/index.php/ESC/article/view/19771/15285.

Ghosh, Arpita, Satyen Kale, and Preston McAfee. 2011. "Who Moderates the Moderators? Crowdsourcing Abuse Detection in User-Generated Content." In *Proceedings of the 12th ACM Conference on Electronic Commerce (EC '11)*, 167–76. New York: ACM.

Gibson, David, Clark Aldrich, and Marc Prensky, eds. 2007. *Games and Simulations in Online Learning: Research and Development Frameworks.* Hershey, PA: Information Science Publishing.

Gitelman, Lisa. 2006. *Always Already New: Media, History, and the Data of Culture.* Cambridge, MA: MIT Press.

GitHub, Inc. 2017. *GitHub.* https://github.com.

Glass, Geof. 2005. *Marginalia.* http://webmarginalia.net/.

Google Inc. 2012. *Google Drive.* https://www.google.com/drive/.

Graff, Gerald. 1987. *Professing Literature: An Institutional History.* Chicago: University of Chicago Press.

_____ 2003. "Introduction: In the Dark All Eggheads Are Gray." In *Clueless in Academe: How Schooling Obscures the Life of the Mind*, 1–16. New Haven, CT: Yale University Press.

Graham, Adam. 2012. "Gamification: Where's the Fun in That?" [blog post]. *Campaign* 43: 47. http://www.campaignlive.co.uk/article/1156994/gamification-wheres-fun-that.

Gregory, Derek. 1994. *Geographical Imaginations.* Oxford: Blackwell.

Grimes, Sara M., and Andrew Feenberg. 2009. "Rationalizing Play: A Critical Theory of Digital Gaming." *The Information Society* 25 (2): 105–18. doi:10.1080/01972240802701643.

Grimshaw, Mark. 2003. *WIKINDX.* http://wikindx.sourceforge.net.

Groh, Fabian. 2012. "Gamification: State of the Art Definition and Utilization." In *Proceedings of the 4th Seminar on Research Trends in Media Informatics (RTMI '12)*, edited by Naim Asaj, et al. Ulm, Germany: Institute of Media Informatics, Ulm University, 39–46. http://hubscher.org/roland/courses/hf765/readings/Groh_2012.pdf.

Gruzd, Anatoliy. 2006–16. *Netlytic.* http://netlytic.org/.

Guédon, Jean-Claude. 2008. "Digitizing and the Meaning of Knowledge." *Academic Matters* (October–November): 23–26. http://www.academicmatters.ca/assets/AM_SEPT'08.pdf.

Guldi, Jo. 2013. "Reinventing the Academic Journal." In *Hacking the Academy: New Approaches to Scholarship and Teaching from Digital Humanities*, edited by Daniel J. Cohen and Tom Scheinfeldt, 19–24. Ann Arbor: University of Michigan Press. doi:10.3998/dh.12172434.0001.001.

Guldi, Jo, and Cora Johnson-Roberson. 2012. *Paper Machines.* metaLAB @ Harvard. http://papermachines.org/.

Habermas, Jürgen. 1991. "Introduction: Preliminary Demarcation of a Type of Bourgeois Public Sphere." In *The Structural Transformation of the Public Sphere*, translated by Thomas Burger with the assistance of Frederick Lawrence, 1–26. Cambridge, MA: MIT Press.

Haft, Jamie. 2012. "Publicly Engaged Scholarship in the Humanities, Arts, and Design." *Animating Democracy*: 1–15. http://imaginingamerica.org/wp-content/uploads/2015/09/JHaft-Trend-Paper.pdf.

Hammond, Adam, and Julian Brooke. 2011–12. *He Do the Police in Different Voices*. http://hedothepolice.org/.

Haraway, Donna. 1990. "A Cyborg Manifesto: Science, Technology, and Socialist Feminism in the Late Twentieth Century." In *Simians, Cyborgs, and Women: The Reinvention of Nature*, 149–81. New York: Routledge.

Hart, Jennefer, Charlene Ridley, Faisal Taher, Corina Sas, and Alan J. Dix. 2008. "Exploring the Facebook Experience: A New Approach to Usability." In *Proceedings of the 5th Nordic Conference on Human-Computer Interaction (NordiCHI08)*, 471–74. New York: ACM.

Hart, William, and Terry Marsh. 2014. "Social Media Research Foundation." In *Encyclopedia of Social Media and Politics*, edited by Kerric Harvey, 3: 1173–74. Thousand Oaks, CA: Sage.

Harvard University Herbarium and University of Massachusetts Boston (UMass-Boston) Biodiversity Informatics Lab. 2010–15. *FilteredPush*. http://wiki.filteredpush.org/wiki/FilteredPush.

Harvard University Library Lab. 2011. *Highbrow*. https://osc.hul.harvard.edu/liblab/projects/highbrow-textual-annotation-browser.

Hayles, N. Katherine. 2007. "Hyper and Deep Attention: The Generational Divide in Cognitive Modes." *Profession*: 187–99.

—————— 2008. *Electronic Literature: New Horizons for the Literary*. Notre Dame, IN: University of Notre Dame Press.

Haystack Group and Massachusetts Institute of Technology (MIT). n.d. *nb*. http://nb.mit.edu/.

Heidegger, Martin. 1982. "The Question Concerning Technology." In *The Question Concerning Technology and Other Questions*, translated with an introduction by William Lovitt, 3–35. New York: Harper Perennial.

Hendry, David G., J.R. Jenkins, and Joseph F. McCarthy. 2006. "Collaborative Bibliography." *Information Processing & Management* 42 (3): 805–25. doi:10.1016/j.ipm.2005.05.007.

Høgenhaug, Peter Steen. 2012. "Gamification and UX: Where Users Win or Lose." *Smashing Magazine.* n.p. http://uxdesign.smashingmagazine. com/2012/04/26/gamification-ux-users-win-lose/.

Holley, Rose. 2010. "Crowdsourcing: How and Why Should Libraries Do It?" *D-Lib Magazine* 16 (3/4): n.p. doi:10.1045/march2010-holley.

Huffman, Steve, and Alexis Ohanian. 2005. *Reddit.* https://www.reddit.com.

Huizinga, Johan. 1949. *Homo Ludens: A Study of the Play-element in Culture.* London: Routledge and Kegan Paul.

Insemtives. 2009–12. *INSEMTIVES.* http://insemtives.eu/.

Introna, Lucas D., and Helen Nissenbaum. 2000. "Shaping the Web: Why the Politics of Search Engines Matters." *The Information Society* 16 (3): 169–85.

Inversini, Alessandro, Rogan Sage, Nigel Williams, and Dimitrios Buhalis. 2015. "The Social Impact of Events in Social Media Conversation." In *Information and Communication Technologies in Tourism 2015*, edited by Iis Tussyadiah and Alessandro Inversini, 283–94. Lugano, Switzerland: Springer International Publishing.

Ittersum, Martine J. van. 2011. "Knowledge Production in the Dutch Republic: The Household Academy of Hugo Grotius." *Journal of the History of Ideas* 72 (4): 523–48. doi:10.1353/jhi.2011.0033.

Jacoby, John James (lead developer). 2009. *BuddyPress.* http://buddypress. org/.

Jagoda, Patrick. 2013. "Gamification and Other Forms of Play." *Boundary 2* 40 (2): 113–44. doi:10.1215/01903659-2151821.

——————— 2014. "Gaming the Humanities." *Differences* 25 (1): 189–215.

Jagodzinski, Cecile M. 2008. "The University Press in North America: A Brief History." *Journal of Scholarly Publishing* 40 (1): 1–20. doi:10.1353/ scp.0.0022.

Jakobsson, Mikael. 2011. "The Achievement Machine: Understanding Xbox 360 Achievements in Gaming Practices." *Game Studies: The International Journal of Computer Game Research* 11 (1): n.p. http://gamestudies. org/1101/articles/jakobsson.

Jankowski, Nicholas W., Andrea Scharnhorst, Clifford Tatum, and Zuotian Tatum. 2013. "Enhancing Scholarly Publications: Developing Hybrid Monographs in the Humanities and Social Sciences." *Scholarly and Research Communication* 4 (1): n.p. http://src-online.ca/index.php/src/article/view/40/123.

Jay, Gregory. 2012. "The Engaged Humanities: Principles and Practices for Public Scholarship and Teaching." *Journal of Community Engagement and Scholarship* 3 (1): 51–63. http://jces.ua.edu/the-engaged-humanities-principles-and-practices-for-public-scholarship-and-teaching/.

Jensen, Matthew. 2012. "Engaging the Learner." *Training and Development* 66 (1): 40.

Jenstad, Janelle, and Kim McLean-Fiander. n.d. "The *MoEML* Gazetteer of Early Modern London." *The Map of Early Modern London*, edited by Janelle Jenstad. Victoria: University of Victoria. http://mapoflondon.uvic.ca/gazetteer_about.htm.

Jessop, Martyn. 2008. "Digital Visualization as a Scholarly Activity." *Digital Scholarship in the Humanities* (formerly *Literary and Linguistic Computing*) 23 (3): 281–93. doi:10.1093/llc/fqn016.

Jing, Tee Wee, Yue Wong Seng, and Raja Kumar Murugesan. 2015. "Learning Outcome Enhancement via Serious Game: Implementing Game-Based Learning Framework in Blended Learning Environment." *5th International Conference on IT Convergence and Security (ICITCS)*, 24–27 August. doi:10.1109/ICITCS.2015.7292992.

Johns, Adrian. 1998. *The Nature of the Book: Print and Knowledge in the Making.* Chicago: University of Chicago Press.

Jones, Steven E. 2009. "Second Life, Video Games, and the Social Text." *PMLA* 124 (1): 264–72.

_____ 2011. "Performing the Social Text: Or, What I Learned from Playing Spore." *Common Knowledge* 17 (2): 283–91. doi:10.1215/0961754X-1187977.

_____ 2013a. *The Emergence of the Digital Humanities.* London and New York: Routledge.

_____ 2013b. "Publications." In *The Emergence of the Digital Humanities*, 147–77. London and New York: Routledge.

Kapp, Karl M. 2012. *The Gamification of Learning and Instruction: Game-Based Methods and Strategies for Training and Education*. San Francisco: Pfeiffer.

Kaps, Jens-Peter. 2003–9. *Document Database*. http://docdb.sourceforge.net.

Kaufer, David S., and Kathleen M. Carley. 1993. "Academia." In *Communication at a Distance: The Influence of Print on Sociocultural Organization and Change*, 341–93. Hillsdale, NJ: Lawrence Earlbaum Associates.

KDE Group at the University of Kassel, DMIR Group at the University of Würzburg, and L3S Research Center, Hannover. 2006. *BibSonomy*. http://www.bibsonomy.org.

Kim, Bohyun. 2012. "Harnessing the Power of Game Dynamics: Why, How to, and How Not to Gamify the Library Experience." *College & Research Libraries News* 73 (8): 465–69.

Kingsley, Danny. 2013. "Build It and They Will Come? Support for Open Access in Australia." *Scholarly and Research Communication* 4 (1): n.p. http://src-online.ca/index.php/src/article/viewFile/39/121.

Kirschenbaum, Matthew. 2012a. "Digital Humanities As/Is a Tactical Term." In *Debates in the Digital Humanities*, edited by Matthew K. Gold, 415–28. Minneapolis: University of Minnesota Press. http://dhdebates.gc.cuny.edu/debates/part/7.

_____. 2012b. "What is Digital Humanities and What's It Doing in English Departments?" In *Debates in the Digital Humanities*, edited by Matthew K. Gold, 3–11. Minneapolis: University of Minnesota Press. http://dhdebates.gc.cuny.edu/debates/text/38.

Kittur, Aniket, and Robert E. Kraut. 2008. "Harnessing the Wisdom of the Crowds in Wikipedia: Quality Through Coordination." In *Proceedings of the 2008 ACM Conference on Computer Supported Cooperative Work (CSCW 08)*, 37–46. New York: ACM.

Kjellberg, Sara. 2010. "I am a Blogging Researcher: Motivations for Blogging in a Scholarly Context." *First Monday* 15 (8): n.p. doi:10.5210/fm.v17i7.3968.

Kopas, Merritt. 2012. *Lim* [video game]. New York: Games for Change. http://www.gamesforchange.org/play/lim/.

Lane, Richard J. 2014. "Innovation through Tradition: New Scholarly Publishing Applications Modelled on Faith-Based Electronic Publishing and Learning Environments." *Scholarly and Research Communication* 5 (4): n.p. http://src-online.ca/index.php/src/article/view/188.

Latour, Bruno. 2009. "A Cautious Prometheus? A Few Steps Towards a Philosophy of Design (with Special Attention to Peter Sloterdijk)." In *Networks of Design: Proceedings of the 2008 Annual International Conference of the Design History Society*, edited by Fiona Hackne, Jonathan Glynne, and Viv Minto, 2–10. Boca Raton, FL: Universal Publishers. 1–13. Cornwall.http://www.bruno-latour.fr/sites/default/files/112–DESIGN-CORNWALL-GB.pdf.

LearningTimes, LLC. 2013. *BadgeOS*. http://badgeos.org/badgestack/.

Lessig, Lawrence. 2004. *Free Culture: How Big Media Uses Technology and the Law to Lock Down Culture and Control Creativity*. New York: Penguin.

Levy, Michelle. 2010. "Austen's Manuscripts and the Publicity of Print." *ELH* 77 (4): 1015–40. https://muse.jhu.edu/journals/elh/v077/77.4.levy.pdf.

Lightman, Harriet, and Ruth N. Reingold. 2005. "A Collaborative Model for Teaching E-Resources: Northwestern University's Graduate Training Day." *Libraries and the Academy* 5 (1): 23–32. doi:10.1353/pla.2005.0008.

Liu, Alan. 2004. *The Laws of Cool: Knowledge Work and the Culture of Information*. Chicago: University of Chicago Press.

_____ 2009. "The End of the End of the Book: Dead Books, Lively Margins, and Social Computing." *Michigan Quarterly Review* 48 (4): 499–520.

_____ 2011. "Friending the Past: The Sense of History and Social Computing." *New Literary History: A Journal of Theory and Interpretation* 42 (1): 1–30. doi:10.1353/nlh.2011.0004.

_____ 2012. "Where is Cultural Criticism in the Digital Humanities?" In *Debates in the Digital Humanities*, edited by Matthew K. Gold, 490–510. Minneapolis: University of Minnesota Press. http://dhdebates.gc.cuny.edu/debates/text/20.

_____ 2013. "From Reading to Social Computing." In *Literary Studies in the Digital Age: An Evolving Anthology*, edited by Kenneth M. Price and Raymond G. Siemens, n.p. New York: MLA Commons. https://dlsanthology.commons.mla.org/from-reading-to-social-computing/.

Liu, Yefeng, Todorka Alexandrova, and Tatsuo Nakajima. 2011. "Gamifying Intelligent Environments." In *Proceedings of the 2011 International ACM Workshop on Ubiquitous Meta User Interfaces (Ubi-MUI '11)*, 7–12. New York: ACM. doi:10.1145/2072652.2072655.

Lorimer, Rowland. 2013. "Libraries, Scholars, and Publishers in Digital Journal and Monograph Publishing." *Scholarly and Research Communication* 4 (1): n.p. http://src-online.ca/index.php/src/article/viewFile/43/117.

_____ 2014. "A Good Idea, a Difficult Reality: Toward a Publisher/Library Open Access Partnership." *Scholarly and Research Communication* 5 (4): n.p. n.p. http://src-online.ca/index.php/src/article/view/180.

Losh, Elizabeth. 2012. "Hacktivism and the Humanities: Programming Protest in the Era of the Digital University." In *Debates in the Digital Humanities*, edited by Matthew K. Gold, 161–86. Minneapolis: University of Minnesota Press. http://dhdebates.gc.cuny.edu/debates/text/32.

Manovich, Lev. 2001. *The Language of New Media.* Cambridge, MA: MIT Press.

_____ 2012. "Trending: The Promises and the Challenges of Big Social Data." In *Debates in the Digital Humanities*, edited by Matthew K. Gold, 460–75. Minneapolis: University of Minnesota Press. http://dhdebates.gc.cuny.edu/debates/text/15.

Manzo, Christina, Geoff Kaufman, Sukdith Punjasthitkul, and Mary Flanagan. 2015. "'By the People, For the People': Assessing the Value of Crowd-sourced, User-Generated Metadata." *Digital Humanities Quarterly* 9 (1): n.p. http://www.digitalhumanities.org/dhq/vol/9/1/000204/000204.html.

Marshall, Catherine C. 1997. "Annotation: From Paper Books to the Digital Library." In *Proceedings of the Second ACM International Conference on Digital Libraries (Digital Libraries '97)*, 131–40. Philadelphia: ACM. doi:10.1145/263690.263806.

Massachusetts General Hospital. 2013. *Domeo.* http://dbmi-icode-01.dbmi.pitt.edu:2020/Domeo/login/auth.

Massachusetts Institute of Technology (MIT). 2008. *Citeline.* http://www.simile-widgets.org/wiki/Citeline.

Maxis. 2008. *Spore* [video game]. Redwood City, CA: Electronic Arts Inc. http://www.spore.com/.

Maxis and The Sims Studio. 2000/2006–. *The Sims* [video game]. Redwood City, CA: Electronic Arts Inc. http://www.thesims.com/en-us.

Maxwell, John W. 2014. "Publishing Education in the 21st Century and the Role of the University." *Journal of Electronic Publishing* 17 (2): n.p. http://quod.lib.umich.edu/j/jep/3336451.0017.205?view=text;rgn=main.

——————— 2015. "Beyond Open Access to Open Publication and Open Scholarship." *Scholarly and Research Communication* 6 (3): n.p. http://src-online.ca/index.php/src/article/view/202.

McCarty, Willard. 2005. *Humanities Computing.* New York: Palgrave Macmillan.

McGann, Jerome. 1991. *The Textual Condition.* Princeton, NJ: Princeton University Press.

——————— 2001. *Radiant Textuality: Literature after the World Wide Web.* New York: Palgrave.

——————— 2005. "Like Leaving The Nile. IVANHOE, A User's Manual." *Literature Compass* 2: 1–27. http://www2.iath.virginia.edu/jjm2f/old/compass.pdf.

——————— 2006. "From Text to Work: Digital Tools and the Emergence of the Social Text." *TEXT: An Interdisciplinary Annual of Textual Studies* 16: 49–62. http://www.jstor.org/stable/30227956.

McGillivray, David, Gayle McPherson, Jennifer Jones, and Alison McCandlish. 2016. "Young People, Digital Media Making and Critical Digital Citizenship." *Leisure Studies* 35 (6): 724–38. doi:10.1080/02614367.2015.1062041.

McGonigal, Jane. 2008. "Engagement Economy: The Future of Massively Scaled Collaboration and Participation." Edited by Jess Hemerly and Lisa Mumbach. Palo Alto, CA: Institute for the Future, Technology Horizons Program. http://www.iftf.org/uploads/media/Engagement_Economy_sm_0.pdf.

——————— 2011. *Reality is Broken: Why Games Make Us Better and How They Can Change the World.* New York: Penguin.

McGregor, Heidi, and Kevin Guthrie. 2015. "Delivering Impact of Scholarly Information: Is Access Enough?" *Journal of Electronic Publishing* 18 (3): n.p.

http://quod.lib.umich.edu/j/jep/3336451.0018.302?view=text;rgn=main.

McKenzie, D. F. 1999. *Bibliography and the Sociology of Texts.* Cambridge: Cambridge University Press.

McKinley, Donelle. 2012. "Practical Management Strategies for Crowdsourcing in Libraries, Archives and Museums." Report for the School of Information Management, Faculty of Commerce and Administration, Victoria University of Wellington (New Zealand): n.p. http://nonprofitcrowd. org/wp-content/uploads/2014/11/McKinley-2012–Crowdsourcing-management-strategies.pdf.

McPherson, Tara. 2012. "Why are the Digital Humanities So White? or Thinking the Histories of Race and Computation." In *Debates in the Digital Humanities*, edited by Matthew K. Gold, 139–60. Minnesota: University of Minnesota Press. http://dhdebates.gc.cuny.edu/debates/text/29.

Meadows, Alice. 2015. "Beyond Open: Expanding Access to Scholarly Content." *Journal of Electronic Publishing* 18 (3): n.p. http://quod.lib.umich.edu/j/jep /3336451.0018.301?view=text;rgn=main.

Meier, Sid. 1991. *Civilization* [video game]. Hunt Valley, MD: MicroProse. http://www.civilization.com.

Mendeley Ltd. 2013. *Mendeley.* http://dev.mendeley.com.

Michel, Jean-Baptiste, Yuan Kui Shen, Aviva Presser Aiden, Adrian Veres, Matthew K. Gray, Google Books Team, Joseph P. Pickett, Dale Hoiberg, Dan Clancy, Peter Norvig, Jon Orwant, Steven Pinker, Martin A. Nowak, and Erez Lieberman Aiden. 2011. "Quantitative Analysis of Culture Using Millions of Digitized Books." *Science* 331 (6014): 176–82. doi:10.1126/ science.1199644.

Mideast Youth. 2016. *CrowdVoice:* Tracking Voices of Protest. http://crowd-voice.org/.

Mojang and Microsoft Studios. 2011. *Minecraft* [video game]. Stockholm: Mojang; Redmond, WA: Microsoft Studios. https://minecraft.net.

Moretti, Franco. 1998. *Atlas of the European Novel, 1800–1900.* London: Verso.

_____ 2005. *Graphs, Maps, Trees: Abstract Models for a Literary History.* London and New York: Verso.

Moyle, Martin, Justin Tonra, and Valerie Wallace. 2011. "Manuscript Transcription by Crowdsourcing: Transcribe Bentham." *Liber Quarterly* 20 (3–4): 347–56. doi:10.18352/lq.7999.

Mozilla Foundation. 2011. *Open Badges.* https://openbadges.org.

Mrva-Montoya, Agata. 2012. "Social Media: New Editing Tools or Weapons of Mass Distraction?" *Journal of Electronic Publishing* 15 (1): 1–24. doi:10.3998/3336451.0015.103.

Multiplayer Online Battle Arena (MOBA) games. 2009–.

Mysirlaki, Sofia, and Fotini Paraskeva. 2012. "Leadership in MMOGs: A Field of Research on Virtual Teams." *Electronic Journal of E-Learning* 10 (2): 223–34.

Nakamura, Lisa. 2009. "Don't Hate the Player, Hate the Game: The Racialization of Labor in World of Warcraft." *Critical Studies in Media Communication* 26 (2): 128–44. doi:10.1080/15295030902860252.

_____ 2013. "'Words With Friends': Socially Networked Reading on Goodreads." *PMLA* 128 (1): 238–43. https://lnakamur.files.wordpress.com/2013/04/nakamura-22words-with-friends22-pmla.pdf.

New Zealand Digital Library Project (University of Waikato). 2005–17. *Greenstone Digital Library Software.* http://www.greenstone.org.

Northwestern University. 2004–13. *WordHoard.* http://wordhoard.northwestern.edu/userman/index.html.

Nowviskie, Bethany. 2012a. "A Digital Boot Camp for Grad Students in the Humanities." *The Chronicle of Higher Education.* Last modified April 29, 2012. http://chronicle.com/article/A-Digital-Boot-Camp-for-Grad/131665.

_____ 2012b. "Evaluating Collaborative Digital Scholarship (or, Where Credit is Due)." *Journal of Digital Humanities* 1 (4): n.p. http://journalofdigitalhumanities.org/1-4/evaluating-collaborative-digital-scholarship-by-bethany-nowviskie/.

O'Donnell, Daniel, Heather Hobma, Sandra Cowan, Gillian Ayers, Jessica Bay, Marinus Swanepoel, Wendy Merkley, Kelaine Devine, Emma Dering, Inge

Genee. 2015. "Aligning Open Access Publication with the Research and Teaching Missions of the Public University: The Case of the Lethbridge Journal Incubator (If 'if's and 'and's were pots and pans)." *Journal of Electronic Publishing* 18 (3): n.p. doi:10.3998/3336451.0018.309. http:// quod.lib.umich.edu/j/jep/3336451.0018.309?view=text;rgn=main.

Open Knowledge Foundation. 2009–12. *AnnotateIt / Annotator.* http:// annotateit.org.

_____ 2011–12. *BibServer.* https://github.com/okfn/bibserver.

_____ 2011–13. *TEXTUS.* http://textusproject.org/.

_____ 2016. *CKAN.* http://ckan.org/.

OpenStreetMap Foundation. n.d. *OpenStreetMap.* https://www.openstreet-map.org.

Oversity Ltd. 2006. *CiteULike.* http://www.citeulike.org.

Ovsiannikov, Ilia A., Michael A. Arbib, and Thomas H. McNeill. 1999. "Annotation Technology." *International Journal of Human-Computer Studies* 50 (4): 329–62. doi:10.1006/ijhc.1999.0247.

Pearce, Nick, Martin Weller, Eileen Scanlon, and Sam Kinsley. 2010. "Digital Scholarship Considered: How New Technologies Could Transform Academic Work." *Education* 16 (1): n.p. http://incducation.ca/ineducation/article/view/44.

Pfister, Damien Smith. 2011. "Networked Expertise in the Era of Many-to-Many Communication: On Wikipedia and Invention." *Social Epistemology: A Journal of Knowledge, Culture and Policy* 25 (3): 217–31. doi:10.1080/0269 1728.2011.578306.

Pinterest, Inc. 2013. *Pinterest.* http://pinterest.com/.

Play the Past. 2010. http://www.playthepast.org.

Polytron Corporation. 2012. *Fez* [video game]. http://fezgame.com/.

Powell, Daniel, Raymond G. Siemens, and William R. Bowen, with Matthew Hiebert and Lindsey Seatter. 2015. "Transformation through Integration: The Renaissance Knowledge Network (ReKN) and a Next Wave of

Scholarly Publication." *Scholarly and Research Communication* 6 (2): n.p. http://src-online.ca/index.php/src/article/view/199.

Powell, Daniel, Raymond G. Siemens, and the INKE Research Group. 2014. "Building Alternative Scholarly Publishing Capacity: The Renaissance Knowledge Network (ReKN) as Digital Production Hub." *Scholarly and Research Communication* 5 (4): n.p. http://src-online.ca/index.php/src/ article/view/183.

Protonotes. 2008. *Protonotes.* http://www.protonotes.com/.

Ramsay, Stephen, and Geoffrey Rockwell. 2012. "Developing Things: Notes Toward an Epistemology of Building in the Digital Humanities." In *Debates in the Digital Humanities,* edited by Matthew K. Gold, 75–84. Minneapolis: University of Minnesota Press. http://dhdebates.gc.cuny.edu/debates/ part/3.

Ratto, Matt. 2011a. "Critical Making: Conceptual and Material Studies in Technology and Social Life." *The Information Society* 27 (4): 252–60. doi:10 .1080/01972243.2011.583819.

_____ 2011b. "Open Design and Critical Making." In *Open Design Now: Why Design Cannot Remain Exclusive,* edited by Bas van Abel, Lucas Evers, Roel Klaassen, and Peter Troxler, n.p. Amsterdam: BIS Publishers. http:// opendesignnow.org/index.php/article/critical-making-matt-ratto/.

Ratto, Matt, and Robert Ree. 2012. "Materializing Information: 3D Printing and Social Change." *First Monday* 17 (7): n.p. doi:10.5210/fm.v17i7.3968.

Ratto, Matt, Sara Ann Wylie, and Kirk Jalbert. 2014. "Introduction to the Special Forum on Critical Making as Research Program." *The Information Society* 30 (2): 85–95. doi:10.1080/01972243.2014.875767.

Real-Time Strategy Games (RTSGs). 1982–.

Ridge, Mia. 2013. "From Tagging to Theorizing: Deepening Engagement with Cultural Heritage through Crowdsourcing." *Curator: The Museum Journal* 56 (4): 435–50. doi:10.1111/cura.12046.

Ritterfeld, Ute, Michael Cody, and Peter Vorderer, eds. 2009. *Serious Games: Mechanisms and Effects.* New York and London: Routledge.

Robinson, Peter. 2010. "Electronic Editions for Everyone." In *Text and Genre in Reconstruction: Effects of Digitization on Ideas, Behaviours, Products and*

Institutions, edited by Willard McCarty, 145–63. Cambridge: Open Book Publishers.

Rockstar Games, Inc. 1997. *Grand Theft Auto* (GTA) [video game]. http://www. rockstargames.com/grandtheftauto/.

Rockwell, Geoffrey. 2003. "Serious Play at Hand: Is Gaming Serious Research in the Humanities?" *TEXT Technology* 2: 89–99.

_____ 2012. "Crowdsourcing the Humanities: Social Research and Collaboration." In *Collaborative Research in the Digital Humanities*, edited by Marilyn Deegan and Willard McCarty, 135–54. Farnham, UK, and Burlington, VT: Ashgate.

Rose, Frank. 2011. *The Art of Immersion: How the Digital Generation is Remaking Hollywood, Madison Avenue, and the Way We Tell Stories*. New York: W.W. Norton.

Rosenzweig, Roy. 2006. "Can History Be Open Source? Wikipedia and the Future of the Past." *Journal of American History* 93 (1): 117–46.

Ross, Anthony, and Nadia Caidi. 2005. "Action and Reaction: Libraries in the Post 9/11 Environment." *Library and Information Science Research* 27 (1): 97–114. doi:10.1016/j.lisr.2004.09.006.

Ross, Stephen, Alex Christie, and Jentery Sayers. 2014. "Expert/Crowdsourcing for the Linked Modernisms Project." *Scholarly and Research Communication* 5 (4): n.p. http://src-online.ca/index.php/src/article/viewFile/186/368.

Roy Rosenzweig Center for History and New Media (George Mason University). 2006–17. *Zotero*. https://www.zotero.org/.

_____ 2007–17. *Omeka*. http://omeka.org.

Ryan, Marie-Laure. 1994. "Immersion vs. Interactivity: Virtual Reality and Literary Theory." *Postmodern Culture* 5 (1): 110–37. doi:10.1353/ sub.1999.0015.

Saklofske, Jon. 2012. "Fluid Layering: Reimagining Digital Literary Archives Through Dynamic, User-generated Content." *Scholarly and Research Communication* 3 (4): n.p. http://src-online.ca/index.php/src/article/ viewFile/70/181.

Saklofske, Jon, and Jake Bruce, with the INKE Research Group. 2013. "Beyond Browsing and Reading: The Open Work of Digital Scholarly Editions." *Scholarly and Research Communication* 4 (3): n.p. http://src-online.ca/index.php/src/article/view/119.

Salen, Katie, and Eric Zimmerman. 2004. *Rules of Play: Game Design Fundamentals.* Cambridge, MA: MIT Press.

San Martin, Patricia Silvana, Paola Caroline Bongiovani, Ana Casali, and Claudia Deco. 2015. "Study on Perspectives Regarding Deposit on Open Access Repositories in the Context of Public Universities in the Central-Eastern Region of Argentina." *Scholarly and Research Communication* 6 (1): n.p. http://src-online.ca/index.php/src/article/view/145.

Schenold, Terry. 2011. "The 'Rattomorphism' of Gamification." *Critical Gaming Project.* https://depts.washington.edu/critgame/wordpress/2011/11/the-rattomorphism-of-gamification/.

Scholars' Lab (University of Virginia Library). 2012. *Prism.* http://prism.scholarslab.org.

Scholz, Trebor, ed. 2013. *Digital Labor: The Internet as Playground and Factory.* New York: Routledge.

Schreibman, Susan, Raymond G. Siemens, and John Unsworth, eds. 2004. *A Companion to Digital Humanities.* Oxford: Blackwell.

Shillingsburg, Peter. 2006. *From Gutenberg to Google: Electronic Representations of Literary Texts.* Cambridge: Cambridge University Press.

Siemens, Lynne. 2009. "It's a Team if You Use 'Reply All': An Exploration of Research Teams in Digital Humanities Environments." *Digital Scholarship in the Humanities* (formerly *Literary and Linguistic Computing*) 24 (2): 225–33. doi:10.1093/llc/fqp009.

Siemens, Raymond G. 2002. "Scholarly Publishing at its Source, and at Present." In *The Credibility of Electronic Publishing: A Report to the Humanities and Social Sciences Federation of Canada,* compiled by Raymond G. Siemens, Michael Best, Elizabeth Grove-White, Alan Burk, James Kerr, Andy Pope, Jean-Claude Guédon, Geoffrey Rockwell, and Lynne Siemens. *TEXT Technology* 11 (1): 1–128.

Siemens, Raymond G., Meagan Timney, Cara Leitch, Corina Koolen, and Alex Garnett, with the ETCL, INKE, and PKP Research Groups. 2012. "Toward Modeling the Social Edition: An Approach to Understanding the Electronic Scholarly Edition in the Context of New and Emerging Social Media." *Digital Scholarship in the Humanities* (formerly *Literary and Linguistic Computing*) 27 (4): 445–61. doi:10.1093/llc/fqs013.

Sinclair, Stéfan, Geoffrey Rockwell, and the Voyant Tools Team. 2012. *Voyant Tools.* http://voyant-tools.org.

Slashdot Media. 2017. *SourceForge.* http://sourceforge.net.

Smith, Martha Nell. 2004. "Electronic Scholarly Editing." In *A Companion to Digital Humanities,* edited by Susan Schreibman, Raymond G. Siemens, and John Unsworth, 306–22. Oxford: Blackwell.

Squire, Kurt. 2008. "Open-Ended Video Games: A Model for Developing Learning for the Interactive Age." In *The Ecology of Games: Connecting Youth, Games, and Learning,* edited by Katie Salen, 167–98. Cambridge, MA: MIT Press.

Stack Exchange Network. 2013. *Stack Overflow.* http://stackoverflow.com.

Stanford Natural Language Processing Group. 2006. *Stanford Named Entity Recognizer (NER).* http://nlp.stanford.edu/software/CRF-NER.html.

Stein, Bob. 2015. "Back to the Future." *Journal of Electronic Publishing* 18 (2): n.p. doi:10.3998/3336451.0018.204.

Stillinger, Jack. 1994. *Coleridge and Textual Instability: The Multiple Versions of the Major Poems.* Oxford: Oxford University Press.

Streeter, Thomas. 2010. "Introduction." In *The Net Effect: Romanticism, Capitalism, and the Internet,* 1–16. New York and London: New York University Press.

StumbleUpon, Inc. 2017. *StumbleUpon.* http://www.stumbleupon.com.

Suits, Bernard. 2005. *The Grasshopper: Games, Life, and Utopia.* Introduction by Thomas Hurka. Peterborough, ON: Broadview Press.

Svensson, Patrik. 2012. "Beyond the Big Tent." In *Debates in the Digital Humanities,* edited by Matthew K. Gold, 36–49. Minneapolis: University of Minnesota Press. http://dhdebates.gc.cuny.edu/debates/text/22.

Tally, Robert T., Jr. 2013. *Spatiality*. London and New York: Routledge.

TAPoR Team. 2015. *TAPoR* (version 3.0). http://www.tapor.ca.

Tejeda, Eddie A. 2008–11. *Digress.it.* http://digress.it/.

Textensor. 2008. *A.nnotate.* http://a.nnotate.com/index.html.

Transliteracies Project (University of California Santa Barbara). 2012. *RoSE.* http://rose.english.ucsb.edu/.

Turner, Fred. 2006. *From Counterculture to Cyberculture: Stewart Brand, the Whole Earth Network, and the Rise of Digital Utopianism.* Chicago: University of Chicago Press.

University of Hamburg. 2009. *CATMA.* http://catma.de.

University of Southampton. 2017. *EPrints.* http://www.eprints.org.

Unsworth, John. 2000. "Scholarly Primitives: What Methods Do Humanities Researchers Have in Common, and How Might Our Tools Reflect This?" Part of a symposium on *Humanities Computing: Formal Methods, Experimental Practice*, sponsored by King's College, London, May 13. http://people. virginia.edu/~jmu2m/Kings.5-00/primitives.html.

Vaidhyanathan, Siva. 2002. "The Content-Provider Paradox: Universities in the Information Ecosystem." *Academe* 88 (5): 34–37. doi:10.2307/40252219.

Vandendorpe, Christian. 2012. "Wikisource and the Scholarly Book." *Scholarly and Research Communication* 3 (4): n.p. http://src-online.ca/src/index. php/src/article/viewFile/58/146.

_____ 2015. "Wikipedia and the Ecosystem of Knowledge." *Scholarly and Research Communication* 6 (3): n.p. http://src-online.ca/index.php/ src/article/view/201.

Van de Sompel, Herbert, Sandy Payette, John Erickson, Carl Lagoze, and Simeon Warner. 2004. "Rethinking Scholarly Communication: Building the System that Scholars Deserve." *D-Lib Magazine* 10 (9): n.p. doi:10.1045/ september2004-vandesompel. http://www.dlib.org/dlib/september04/ vandesompel/09vandesompel.html.

Van House, Nancy A. 2003. "Digital Libraries and Collaborative Knowledge Construction." In *Digital Library Use: Social Practice in Design and Evaluation,* edited by Ann Peterson Bishop, Nancy A. Van House, and Barbara P. Buttenfield, 271–95. Cambridge, MA: MIT Press.

Van Staalduinen, Jan-Paul, and Sara de Freitas. 2011. "A Game-Based Learning Framework: Linking Game Design and Learning Outcomes." In *Learning to Play: Exploring the Future of Education with Video Games*, edited by Myint Swe Khine, 29–54. New York: Peter Lang.

Vetch, Paul. 2010. "From Edition to Experience: Feeling the Way towards User-Focused Interfaces." In *Electronic Publishing: Politics and Pragmatics*, edited by Gabriel Egan, 171–84. New Technologies in Medieval and Renaissance Studies 2. Tempe, AZ: Iter Inc., in collaboration with the Arizona Center for Medieval and Renaissance Studies.

Vision Critical Communications. 2013. *DiscoverText*. http://discovertext.com.

Walsh, Brandon, Claire Maiers, Gwen Nally, Jeremy Boggs, and Praxis Program Team. 2014. "Crowdsourcing Individual Interpretations: Between Microtasking and Multitasking." *Digital Scholarship in the Humanities* (formerly *Literary and Linguistic Computing*) 29 (3): 379–86. doi:10.1093/llc/fqu030.

Wark, McKenzie. 2007. *Gamer Theory*. Cambridge, MA: Harvard University Press.

Wasik, Bill. 2009. *And Then There's This: How Stories Live and Die in Viral Culture*. New York: Viking.

Werbach, Kevin. 2014. "(Re)Defining Gamification: A Process Approach." In *Proceedings of the 9th International Conference on Persuasive Technology (PERSUASIVE 2014), Lecture Notes in Computer Science* 8462: 266–72. doi:10.1007/978-3-319-07127-5_23.

Westphal, Bertrand. 2011. *Geocriticism: Real and Fictional Spaces*. Translated by Robert T. Tally, Jr. New York: Palgrave Macmillan.

Whaley, Dan. 2011. *Hypothes.is*. https://hypothes.is/.

Wick, Marc (founder), and Christophe Boutreux (developer). *GeoNames*. Männedorf, Switzerland: Unxos GmbH. http://www.geonames.org.

Williams, George H. 2012. "Disability, Universal Design, and the Digital Humanities." In *Debates in the Digital Humanities*, edited by Matthew K. Gold, 202–12. Minneapolis: University of Minnesota Press.

Wolfe, Joanna. 2002. "Annotation Technologies: A Software and Research Review." *Computers and Composition* 19 (4): 471–97. doi:10.1016/S8755-4615(02)00144-5.

Wrisley, David J., and the team at the American University of Beirut. 2016. *Linguistic Landscapes of Beirut* (formerly *Mapping Language Contact in Beirut*). http://llb.djwrisley.com/.

Wylie, Sara Ann, Kirk Jalbert, Shannon Dosemagen, and Matt Ratto. 2014. "Institutions for Civic Technoscience: How Critical Making is Transforming Environmental Research." *The Information Society* 30 (2): 116–26. doi:10.1080/01972243.2014.875767.

Yahoo Inc. 2005–17. *Flickr*. http://www.flickr.com/.

Zacharias, Robert. 2011. "The Death of the Graduate Student (and the Birth of the HQP)." *English Studies in Canada* 37 (1): 4–8. https://ejournals.library.ualberta.ca/index.php/ESC/article/view/25197/18694.

Zichermann, Gabe, and Christopher Cunningham. 2011. *Gamification By Design: Implementing Game Mechanics in Web and Mobile Apps.* Sebastopol, CA: O'Reilly Media.

Zurb. 2011–17. *Bounce*. http://www.bounceapp.com/.

Zynga Inc. 2009. *FarmVille* [video game]. Facebook and HTML 5. http://company.zynga.com/games/farmville.

Contributors

Alyssa Arbuckle is the Associate Director of the Electronic Textual Cultures Lab at the University of Victoria. She also works with the Implementing New Knowledge Environments Partnership and assists with the coordination of the Digital Humanities Summer Institute. Alyssa is currently pursuing an interdisciplinary PhD at the University of Victoria, studying open social scholarship and its implementation.

Nina Belojevic is a User Experience Strategist at DDB Canada / Tribal Worldwide. In this role she develops strategies and road maps for digital client initiatives, creates user experiences, and works closely with creative teams, designers, developers, and project managers. Nina holds an MA in English with a specialty in Digital Humanities from the University of Victoria.

Tracey El Hajj is a PhD student in the English Department at the University of Victoria. She completed her MA at the American University of Beirut, where she developed a social networking tool that serves literary purposes. Her research interests now include the application of physical computing practices to postwar and postmodern fiction, as well as programming for humanists.

Randa El Khatib is a PhD candidate in the English Department at the University of Victoria. Her research focuses on how space is represented in fictional and allegorical settings of the English Renaissance. Randa is the Special Projects Coordinator at the Electronic Textual Cultures Lab, the Project Manager of the TopoText team (American University of Beirut), and an Alliance of Digital Humanities Organizations Communications Fellow.

Aaron Mauro is Assistant Professor of Digital Humanities and English at Penn State Erie, The Behrend College. He is the Director of the Penn State Digital Humanities Lab, and his research focus includes digital culture, computational text analysis, and scholarly communication. Previously, Aaron was a Postdoctoral Fellow at the Electronic Textual Cultures Lab.

Daniel Powell is a Marie Skłowdowska-Curie Fellow based in the Department of Digital Humanities at King's College London, and affiliated with the Department of English at the University of Victoria. Daniel's research interests include the digital humanities, social knowledge creation, scholarly communications, media archaeology, graduate education in the humanities, cyberinfrastructure, and early modern culture.

ISBN 978-0-86698-739-4 (online) ISBN 978-0-86698-583-3 (print)
New Technologies in Medieval and Renaissance Studies 7 (2017) 265–266

Lindsey Seatter is a PhD candidate studying the British Romantic period and Digital Humanities at the University of Victoria. Her SSHRC-funded dissertation research focuses on exploring the patterns across Jane Austen's print and manuscript work, the evolution of the novel, and reader engagement with narrative practices. Lindsey is also a Graduate Research Assistant at the Electronic Textual Cultures Lab.

Raymond G. Siemens is a Distinguished Professor in the Faculty of Humanities at the University of Victoria, in English with cross appointment in Computer Science. A former Canada Research Chair in Humanities Computing. Raymond is the Director of the Implementing New Knowledge Environments Partnership, the Digital Humanities Summer Institute, and the Electronic Textual Cultures Lab. His research focuses on open social scholarship, textual-editorial intervention, online publishing, and digital humanities communities and teams.